For Reference

Not to be taken from this room

W9-AXJ-756

DOG ANATOMY

PETER C. GOODY   B.Sc. M.Sc(Ed). Ph.D.
Former Lecturer in Anatomy, The Royal Veterinary College, London

Dog Anatomy

A PICTORIAL APPROACH TO CANINE STRUCTURE

J. A. ALLEN : LONDON

BOCA RATON PUBLIC LIBRARY
BOCA RATON, FLORIDA

British Library Cataloguing-in-Publication Data.
A catalogue record for this book is available from the
British Library.

ISBN 0.85131.636.0.

© Peter C. Goody, 1997

No part of this book may be reproduced, stored in a retrieval
system, or transmitted, in any form or by any means, electronic,
mechanical, photocopying, recording or otherwise, without the
prior permission of the publisher. All rights reserved.

First published in Great Britain 1997
Reprinted 2002

J.A.Allen
Clerkenwell House
Clerkenwell Green
London EC1R 0HT

J.A.Allen is an imprint of Robert Hale Limited

Typeset by Setrite Typesetters Ltd., Hong Kong
Printed by Midas Printing International Ltd.

PREFACE

Compared with many other areas of scientific enquiry, gross anatomy (that aspect of anatomy that is readily appreciated with the unaided eye) is one in which little radically new or different information is discovered. Consequently the gross structure of the dog is represented as a well established body of knowledge that has been described and illustrated in considerable detail by a number of authors in the past. However, unless you happen to be a veterinary student, it is unlikely that you will be able to study actual anatomical specimens and so much of this detail will be unavailable to you. Because of this I feel that there is a need to present some of the essential aspects of canine anatomy in a considerably less detailed manner. In producing this book my aim has therefore been to set out some of this established body of knowledge in a series of drawings — my viewpoint being that anatomy is essentially a visual topic in which descriptions without the necessary pictures are wholly inadequate for a real understanding. The drawings are primarily intended to answer questions relating to appearance, size, shape, and position, but also relationships — the questions involved in topographical anatomy.

Returning to the fact that most of you will undoubtedly never see the inside of a dog, I have placed special emphasis on the relationship of structures inside the body to features on the surface. Things that can be felt through the skin — bones, muscles, blood vessels, and so on, form a most important component of many of the drawings. Other structures such as the organs within the chest (eg. heart and lungs) and abdomen (eg. stomach, intestines and kidneys) which cannot be felt from the surface, are nevertheless considered with particular reference to their position in relation to the body surface. In so doing I want to be able to help you 'visualize' what is inside your dog so that you can look at your pet and have a reasonable idea of its internal anatomy.

The illustrations forming the basis of the book I drew myself. They are not intended to be exact 'photographic' representations, my artistic abilities do not extend to that. My illustrative method tends to be diagrammatic to a considerable extent producing simplified drawings intended to explain rather than represent 'real' appearance. I hope that you will be able to visualize the interior of the dog from them. A number of my drawings are modified from or based on illustrations in other books.

Where this is the case due acknowledgement is given in the accompanying legend. All of the drawings are extensively labelled and annotated in accompanying legends, so that illustrations, legends and text contain the essentials of canine anatomy in a readily accessible form.

The text of the book I have deliberately phrased in general terms, it is not intended to refer to any specific breed. On the other hand, many of the drawings are based on the skeleton and body contour of a greyhound. Should you have a greyhound as a pet, or be familiar with short-haired, long-legged breeds, then no doubt you will be able to relate immediately to my drawings. However, should your fancy be a long-haired and/ or short-legged breed then you will undoubtedly have to work a little harder in relating my drawings to your dog! To try and include breed variations would have been beyond the aims of the book. Despite this I hope that you will still look upon the drawings as simply those of a dog.

A further feature of the text requiring mention is the 'naming of parts'. Should any of you have looked at an anatomical treatise of any sort you might well have been exasperated by the terminology that is often used. All those long Latin words can be distinctly off-putting! Thus I have 'anglicized' as many of the anatomical terms as is feasible, in so doing I hope that some barriers to your appreciation of the subject will have been removed. For those of you who might want a more detailed coverage of the anatomy, I have included a short bibliography at the end of the book.

I would like to take this opportunity to thank my publisher Mr. J. A. Allen for kindly agreeing to publish the book for me; my brother John for putting the numbering on my drawings; and Dr. Roy Batt MRCVS for reading through and commenting on the manuscript.

Finally I shall be satisfied should the book encourage you in some small way to consider more closely how your dog is constructed and how it functions, and can only reiterate what I suggested in the introduction to a similar book on Horse Anatomy I produced some years ago — 'since the book is intended to relate all internal structures to specific points or areas on the surface, I think you will agree with me when I suggest that probably the best way to use the book would be in conjunction with a live (dog). You could then verify for yourself the facts illustrated in the drawings'.

PETER GOODY

TABLE OF CONTENTS AND
LIST OF ILLUSTRATIONS

x

INTRODUCTION

The gross structure of the dog is the main theme within the following pages. However, it is often difficult to describe many basic anatomical structures and explain their functions without making some brief mention of the building materials which make them up. I'm sure that you will all be familiar with the idea that the body is made up of cells. Although conforming to a basic general pattern cells are specialized to perform different functions. Furthermore such specialized cell types are not randomly scattered throughout the body, they are collected together by function into tissues. A **tissue** is therefore a collection of like cells which function together to perform a special activity.

As well as cells of the particular functional type, a tissue also contains a variable quantity of fluid both bathing the cells themselves and conveyed in channels between them. Raw materials such as oxygen and food passing to cells, and waste products passing from them, are carried in this **tissue (intercellular) fluid**. A third component which may be present in a tissue is **intercellular material** produced by the cells and located in the tissue fluid. Any tissue examined might therefore consist of three components − cells, fluid and intercellular materials − in variable quantities.

Depending upon their structure and function the body contains tissues of four basic types. Although some tissues may appear to be highly modified they can still be related to one of these four.

Firstly, **epithelial tissue** is a simple tissue type forming covering or lining layers whose main function is probably protective. Cells making up the tissue are closely packed together so that the amount of intercellular material is insignificant. It is found covering the outside of the body in the skin, but also inside the body where it lines the digestive, respiratory, urinary and genital tracts, as well as the blood vessels and the heart. Although resistance to penetration is an important element of protection, subsidiary functions that some epithelial tissue may be called upon to perform are those of absorption and secretion as in the gut lining. Most of the glandular structures in the body in fact are formed from modified epithelium. A further function of certain epithelia is to induce movement of substances in contact with them through the rhythmic beating of tiny hair-like projections from the surface of the epithelial cells (cilia). Such ciliary action is particularly important in, for instance, moving mucus along the nasal cavity and trachea.

Secondly, **muscular tissue** consisting of cells capable of shortening in length and therefore concerned in the main with body movements and maintaining body posture, but also with moving substances such as food, blood and sperm through the body. Three different types of muscle tissue are therefore present in the body − voluntary (striated), involuntary (smooth) and cardiac.

Thirdly, **nervous tissue** based on cells able to conduct 'information' enabling a dog to be aware of what is going on both outside and inside its body, and to make any changes that might be necessary to cope with the varying conditions.

Finally, **connective tissue**, the most abundant body tissue playing a general role in support, extending through and around the other three tissues, supporting, connecting and binding them together. So, for instance, muscle cells making up muscle tissue can have no effect unless they are attached through the medium of connective tissue. Connective tissues contain a large amount of intercellular material in addition to their cells unlike epithelia in which intercellular material is insignificant in amount. In fact the bulk of connective tissue is made up of intercellular material, the cells themselves being fairly widely scattered.

Various categories of connective tissue are recognized based on the degree of fluidity of the intercellular material. At one extreme **blood** is a fluid connective tissue; blood plasma is the intercellular fluid in which blood cells are suspended. At the other extreme of viscosity lies **bone**, a connective tissue in which the intercellular material is solid and crystalline. Between these two extremes all grades of connective tissue occur. Thus there is **loose connective tissue** in which the intercellular material is soft and jelly-like, and the cell products such as fibres are randomly scattered. This is found for instance in the superficial fascia beneath the skin. **Dense connective tissue** has a thickened intercellular matrix in which the fibres are more regularly arranged, and is found in tendons and ligaments. **Cartilage**, in which the intercellular material is stiffened considerably but still flexible, is found in the skeleton of the puppy and in many joints of the adult.

Cells in looser forms of connective tissue may also perform other functions. They may be protective, having the ability to engulf ('swallow') bacteria and cell debris, or to produce antibodies in the defence mechanisms of the body against disease. In superficial fascia beneath the skin connective tissue cells may also store fat.

Although these four basic tissues are distributed throughout the body they can in the main only be recognized with the aid of a microscope. Gross examination of the body with the naked eye shows it to be made up of structurally distinct parts or organs, each performing definite items of work. An **organ** is in fact formed from an intermingling of various types of tissue which have a highly organized structural relationship with one another, although one tissue may be dominant relating to the major function of the particular organ in question. In a clearly recognizable organ such as a kidney all four basic tissues are represented: epithelial tissue forms excretory urinary tubules and lines the permeating renal blood vessels; connective tissue is present as blood, as fibrous and elastic connective tissue making up the bulk of blood vessel walls, as loose connective tissue binding tubules and blood vessels together, and as a dense fibrous capsule around the outer perimeter of the organ; muscular tissue is present in the walls of blood vessels; nervous tissue runs through the organ innervating the muscle in the blood vessel walls. All of these components are organized into a system of interrelated structures whose combined action is the elimination of unwanted chemical substances from the body and the regulation of its water content.

Just as cells and tissues cannot operate in isolation, organs too are interrelated to form **organ systems** in which several organs together perform a specified function. The body is therefore composed of a number of organ systems.

The skin and its appendages comprises the **integumentary system** enclosing all other parts of the body keeping them together. This system *per se* receives little attention in the drawings. But, since we are primarily concerned with the relationship of internal structures to the surface of the body, surface views of the dog form possibly the most important drawings in the book. In view of this the first two figures and the final three are solely devoted to the surface of the body, while many of the remaining figures have surface views of some sort incorporated in them.

The bulk of the animal is formed from the **musculoskeletal system**. This is responsible for support and locomotion and is considered in terms of a **skeletal system** of bones and cartilages providing a mobile framework which gives support and protection to the softer structures, and a **muscular system** providing motile forces. Since this system forms the bulk of the animal giving it shape and form and, since most of the palpable surface features are musculoskeletal, it is given detailed consideration in figures 3 through 18. Figure 19 attempts in a diagrammatic fashion to draw together some of the biomechanical aspects of the two systems.

Primarily located in the body cavities of the thorax, abdomen and pelvis are the major organ systems normally thought of in terms of the 'internal viscera'. The **digestive system** is for obtaining, swallowing and breaking down food into simple substances which can be absorbed into the body. The **respiratory system** is for taking in oxygen and for giving out carbon dioxide from the lungs. The **urogenital system** is responsible for excretion and reproduction and is subdivided into a **urinary system** for removal of waste chemical substances from the body, and a **genital system** for the propagation of the species. These four systems are all considered in some detail in figures 20 through 29 especially in terms of their relationships with the surface of the body. In this consideration two figures (25 and 26) concentrate specifically on the head an especially complex region. These drawings are primarily concerned with visualizing internal structures in relation to the surface since much is not palpable. During this consideration we will look at certain aspects of the organs of special sensation (nose, eyes, ears), components of the **sensory system** responsible for gathering information about the external environment.

Following on from this consideration of viscera we turn our attention to the more diffuse and all pervading systems linking the viscera together. Firstly, the **circulatory system** for circulating the fluid content of the body. This is subdivided into a **blood (cardiovascular) system** of heart and blood vessels circulating blood and ensuring the mixing of all body fluids, and a **lymphatic system** of lymph vessels and nodes returning excess tissue fluid to the general circulation. These systems are considered in figures 30 through 32. Although the lymph nodes are an integral component of the lymphatic system they also include tissue of the **lymphoid (immune) system**. Other grossly recognizable components of the lymphoid system include organs such as the bone marrow, thymus, spleen and tonsils. Secondly, the **nervous system** for communicating information between parts of the body and for governing the activities of the various organs is illustrated in figures 33 through 35.

A final selection of drawings (figures 36 through 38) are specifically of the surface anatomy of the dog and attempt to bring together much of the foregoing information. The various views of the surface attempt to portray most of the structures that are related to the body surface, either palpably or through visualization by surface projection.

As well as some preliminary remarks about the 'building materials' making up the body, some mention is required concerning the descriptive terms that will be used in the text to indicate the position and direction of body parts.

In a normal standing position our dog has a lower or **ventral** surface directed towards the ground, and an opposite upper or **dorsal** surface: the head end of the body is **cranial**, the tail end is **caudal**. The relation of parts in these directions are named accordingly; eg. the neck is cranial to the thorax; the abdomen is caudal to the thorax. Within the head itself the directional term cranial is inapplicable and is replaced by the term **rostral** (towards the muzzle): thus the eyes are rostral to the ears, and so on. Lower down the limbs in the paws, the front or cranial surface is often termed the dorsal surface, while the rear surface may be termed the ventral surface or more specifically the **palmar surface** in the forepaw and the **plantar surface** in the hindpaw.

The terms **lateral** and **medial** are used in the head, neck and trunk to refer to structures or positions away from or closer to the midline – something in the midline is **median**. In the limbs the innerside is medial the outer side is lateral. **Proximal** and **distal** are terms used particularly in the limbs: proximal referring to the upper end close to the body and distal to the lower parts of the limbs and paws.

When we begin to consider muscle action several descriptive terms will be used to explain joint movements. Thus **flexion** or folding of a joint involves movement of one bone in relation to another to reduce the angle between the two: **extension** involves an increase in this angle. **Adduction** and **abduction** refer respectively to the movement of a part towards or away from the median plane and are normally used in movement of a limb in relation to the body. In this context of total limb movement the terms **protraction** and **retraction** will be used: a limb is protracted when it is swung forwards in relation to the body; retraction is backward movement of the limb relative to the body. But remember, with the paw of a protracted limb placed on the ground limb retraction will move the body forwards over the limb, the power thrust of normal locomotion.

Finally should I use the terms right and left, these are determined in relation to the dog not you as the observer – important when the dog is on its back.

1

SURFACE FEATURES OF
THE DOG

On this first drawing of a dog in a normal standing position a number of surface features are labelled. Many of these are 'landmarks' (prominences or depressions of the body contour) and denote the presence and position of underlying structures. I'm sure that some of you will already be familiar with many of these 'points'. Accompanying the main drawing is an enlarged sketch of the head, while the surface of the body is also illustrated from different viewpoints in a number of drawings later in the book. All of these drawings may be consulted for your initial examination. Only in short-haired breeds, however, will a simple visual inspection reveal many surface features. Your examination should therefore be made with hands as well as eyes when the points are obscured by a thicker coat.

The recognizable surface features shown in this drawing are related to several of the organ systems described in the introduction. Thus the skeletal system, the muscular system, the blood and lymphatic systems, the digestive system, the respiratory system, and the reproductive system, are all represented. Nevertheless, the most pronounced and numerous landmarks are those produced by muscles and bones. Many of the lumps and bumps which you can see or feel through the skin are parts of bones which are lying just beneath it in a subcutaneous position, or at most are only covered by thin layers of muscle. In several of the surface views later in the book I have made a special point of highlighting the position of these palpable subcutaneous bony features because of their value as landmarks to the skeleton generally. If you were now to compare this surface drawing with the illustration of the skeleton (fig 3) you will no doubt gain some idea concerning the overall position of the skeleton in the body.

A number of muscles may also be felt through the skin, and some produce raised contours on its surface. These muscles are labelled in this illustration and by comparing it with the first three or four muscle drawings you will also obtain some idea about muscle position. Other soft structures such as blood vessels and nerves are not so apparent, for it is only in a few places that they are superficial in position and so either visible or palpable. It will be obvious to you that these delicate structures are protected from damage to some extent by being routed through areas where they will not be exposed to external trauma. Consequently they are buried more deeply within the body where they are afforded greater protection. In this illustration, of a short-haired dog, the raised outline of the saphenous vein is visible crossing the outer surface of the lower leg; while the cephalic vein is apparent running up the front of the forearm.

Surface features are either seen or felt through the **skin** which covers the entire body. The dominant component of skin is a layer of epithelial tissue of variable thickness, continous with the membrane lining the gut at mouth and anus, lining the respiratory tract at the external nostrils, and lining the urogenital tract at the vulva or the external urethral opening at the tip of the penis. Skin, however, is an organ since it contains various other tissues which are structurally and functionally interrelated with the epithelium. In fact it is probably the largest organ in the body! Thus skin and its derivatives, such as hair, claws, pads, glands, and sensory receptors, make up what is known as the integumentary system or common integument. It forms a protective limiting boundary between a dog and its surroundings, and is to a certain extent responsible for imparting and maintaining its shape. The epithelial layer is only the outermost component, the **epidermis**, beneath which lies the **dermis**, a layer of dense connective tissue rich in collagen and elastic fibres. Firmly attached to the epidermis outside and to deeper tissues inside, the dermis anchors the epidermis in position and imparts strength to the skin. It is also the pathway for raw materials such as food and oxygen to reach epidermal cells since blood vessels and nerves to the skin end within it.

The dermis is joined with underlying structures by quite loose connective tissue, the **superficial fascia**. It is this layer which allows the skin to move in relation to deeper structures to considerable degrees. Such a capacity enhances the protective properties of skin enabling it to withstand the buffetings of everyday life without necessarily being ruptured. In many parts of the body this tissue is infiltrated by variable amounts of fat which not only act in an insulating capacity but also in protection against physical damage. Further enhancement of the protective capacities of skin is afforded by the epidermis being multi-layered. The deepest epidermal layer, next to the dermis, is a germinal layer in which cellular division continually adds new cells to the epidermis. As they are gradually pushed towards the surface by continued cell multiplication beneath them, they get further away from their nutrient source in the dermis. Consequently their vitality decreases, they become hardened and they finally die. The outermost epidermal layers, formed of dying and dead cells, provide protection against injury and the entrance of disease producing organisms, as well as 'waterproofing' the body. The dead surface cells are eventually cast off.

Hairs are epidermal derivatives and you will all be familiar with the enormous range of colour, length, thickness, and texture in the various breeds. Hair tends to be more sparsely distributed where skin is thin, and is absent from the moist nasal plane of the nose, the footpads, the navel and the teats. Each individual hair is a non-living structure produced from a hair follicle embedded in the dermis but lined by the germinative epidermal layer. Although hair is obviously protective its other function is probably to help regulate temperature. It can do this because it can trap a layer of air against the skin surface which acts as an insulator. In cold weather hairs tend to rise forming a thicker insulating blanket cutting down heat loss to the surrounding environment, while in warm weather the coat lies flatter reducing the thickness of the blanket and allowing heat to be dissipated more readily. Such hair movements are produced to some extent by the action of cutaneous muscle sheets in the superficial fascia but are mainly a response to reflex action in minute muscles attached to each hair root in the dermis.

Certain special facial hairs, in the form of long, stiff *tactile hairs* are important for sensory perception. These may be solitary or grouped and are found in various places on the face — on the lips, muzzle, cheeks, throat, and above the eyes — and nerve fibres wrapped around their roots record 'touch' information whenever the hairs are moved. Numerous other nerve endings lie in the dermal layer which means that the skin can act as a receptor organ for several different types of stimulus such as temperature, pain, touch and pressure.

Surface features of head and neck
1 Nasal plane (pigmented hairless skin). 2 External nostril (leading into nasal vestibule surrounded by nasal cartilages). 3–5 Lips (surrounding oral fissure — mouth opening). 3 Upper lip (supporting superior labial tactile sensory hairs). 4 Lower lip. 5 Commissure of lips at angle of mouth. 6 Foreface. 7 Muzzle. 8 Prominence of chin (mentum supporting tactile sensory mental hairs). 9–12 Tactile sensory hairs of face. 9 Supraorbital sensory hairs. 10 Zygomatic sensory hairs. 11 Buccal sensory hairs. 12 Intermandibular sensory hairs. 13 Cheek (based on buccinator muscle). 14 Eyeball (situated in orbit and protected by a bony orbital rim). 15–16 Eyelids (surrounding palpebral fissure). 15 Upper eyelid supporting cilia (eyelashes). 16 Lower eyelid.

17 Stop. 18 Forehead. 19 Pinna (visible part of external ear based on auricular cartilage). 20 Marginal cutaneous pouch of helix. 21 Tragi (prominent hairs at opening into ear canal). 22 Tongue. 23–25 Teeth of lower dental arch. 23 Incisor teeth. 24 Canine tooth (eye-tooth). 25 Lower carnassial (shearing) tooth (molar 1). 26 Crest of neck. 27 Throat. 28 Jugular groove (containing external jugular vein). 29 Jugular fossa (triangular depression at base of neck).

Surface features of trunk and tail
30 Breast (based on pectoral muscles). 31 Withers (interscapular region). 32 Brisket (chest). 33 Back (dorsal region). 34 Umbilicus (navel — hairless scar denoting point of entry and exit of blood vessels in foetus). 35 Belly. 36 Flank. 37 Fold of flank (running onto thigh proximal to stifle joint). 38 Loins (lumbar region). 39 Croup (sacral region). 40 Rump (hindquarters). 41 Prepuce (sheath covering and protecting glans penis). 42 Ischiorectal fossa (depression lateral to root of tail and normally fat filled). 43 Root of tail (set-on of tail). 44 Tail.

Surface features and regions of limbs
45 Shoulder. 46 Arm (brachium or upper arm). 47 Axilla (armpit). 48 Forearm (antebrachium). 49–51 Forepaw. 49 Carpus (wrist — a topographical region based on carpal bones and carpal joints). 50 Metacarpus (front pastern based on metacarpal bones). 51 Digits (toes based on phalangeal bones). 52 Thigh (upper thigh). 53 Calf (based on gastrocnemius muscle). 54 Popliteal fossa (caudal to stifle joint containing popliteal lymph node). 55 Shank (leg, crus or lower thigh). 56–57 Hindpaw. 56 Hock (tarsus or ankle — a topographical region based on tarsal bones and joints). 57 Metatarsus (rear pastern based on metatarsal bones). 58–60 Pads of paws. 58

Carpal (stopper) pad. 59 Metacarpal pad of forepaw, metatarsal pad of hindpaw. 60 Digital pads, 61 Claw (unguis — capping ungual process of distal phalanx). 62 Wall of claw. 63 Sole of claw. 64 Interdigital space.

Bony landmarks of head, neck and trunk
65 Zygomatic (supraorbital) process of frontal bone. 66 Orbital ligament (joining frontal bone and zygomatic arch completing orbital rim). 67 External sagittal crest. 68 External occipital protuberance (occiput). 69 Zygomatic arch (bridge of bone connecting face and cranium below eye). 70 Body of mandible (lower jaw). 71 Thyroid cartilage (forming 'laryngeal prominence' of voice box). 72 Wing of atlas (transverse process of 1st cervical vertebra). 73 Costal arch (fused costal cartilages of ribs 10–12 attached to costal cartilage of rib 9). 74 Rib 13 (last or floating rib normally attached by fibrous tissue with costal arch).

Bony landmarks of limbs
75–77 Scapula (shoulder blade). 75 Dorsal (vertebral) border of scapula. 76 Spine of scapula. 77 Acromion process of scapula. 78 Point of shoulder (greater tubercle of humerus). 79 Point of elbow (olecranon process of ulna). 80 Lateral styloid process of ulna. 81 Medial styloid process of radius. 82 Accessory carpal bone. 83–85 Hip (pelvic) bone. 83 Sacral tuberosity of ilium (point of croup — cranial dorsal iliac spine). 84 Coxal tuberosity of ilium (point of haunch — cranial ventral iliac spine). 85 Ischiatic tuberosity of ischium (point of buttock or seat bone). 86 Greater trochanter of femur (point of hip). 87 Patella ('knee cap' — sesamoid bone in tendon of insertion of quadriceps femoris muscle). 88–89 Tibia. 88 Tuberosity of tibia (insertion for patellar tendon). 89 Medial malleolus of tibia. 90 Lateral malleolus of fibula. 91 Calcaneal tuberosity (point of hock — area

of attachment for common calcaneal tendon from calf muscles).

Position of joints
92 Jaw (temporomandibular) joint. 93 Shoulder (scapulohumeral) joint. 94 Elbow (cubital) joint. 95 Metacarpophalangeal joints of forepaw and metatarsophalangeal joints of hindpaw. 96 Proximal interphalangeal joint. 97 Distal interphalangeal joint. 98 Hip (coxofemoral) joint. 99 Stifle (knee) joint. 100 Talocrural (crurotarsal) joint.

Muscles producing identifiable contours
101–102 Jaw closure muscles. 101 Temporal muscle. 102 Masseter muscle. 103 Epaxial musculature (extending whole length of neck, trunk and tail). 104 Brachiocephalic muscle (major limb protractor). 105 Latissimus dorsi muscle (major limb retractor). 106 Long head of triceps brachii muscle (forming caudal border [tricipital margin] of arm). 107 Sartorius muscle (forming cranial border of thigh). 108 Extensor muscles of carpus and digits (craniolateral muscle mass of forearm). 109 Flexor muscles of carpus and digits (caudomedial muscle mass of forearm). 110 Tendon of ulnar carpal flexor muscle (taut cord attached to accessory carpal bone). 111 Tendons of deep and superficial digital flexor muscles. 112 'Hamstring' muscles (biceps femoris, semitendinosus and semimembranosus). 113 Patellar tendon (continuation onto tibia of quadriceps femoris tendon and containing patella). 114 Common calcaneal tendon (aggregate of tendons attached to point of hock including Achilles' tendon from gastrocnemius and tarsal tendons from hamstrings).

Blood vessels
115 Cephalic vein of forelimb. 116 Lateral saphenous vein of hindlimb.

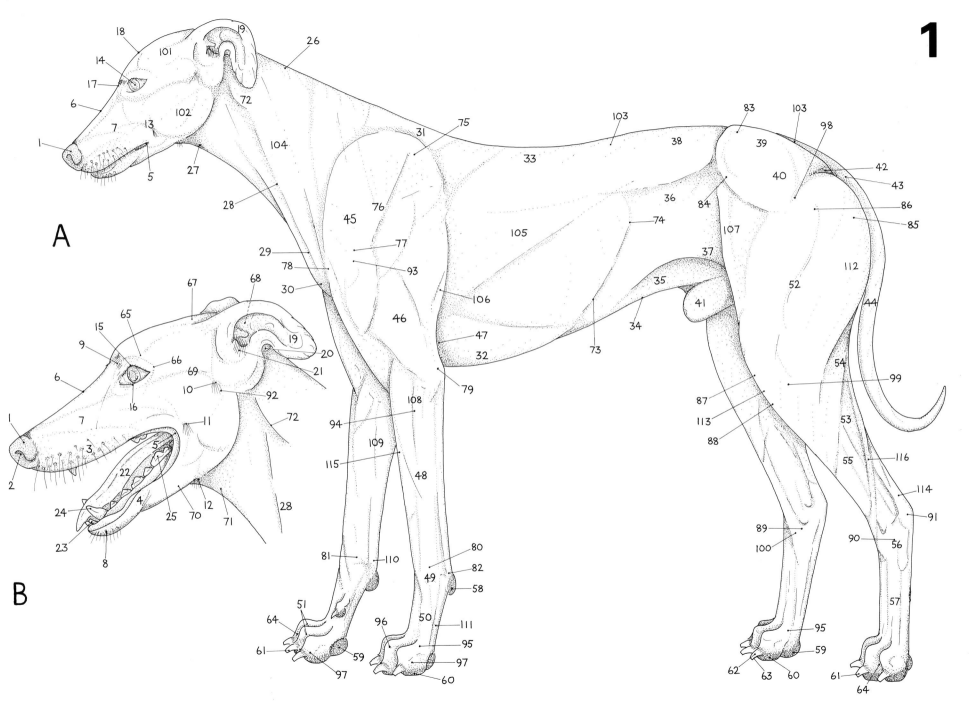

A

B

1

2

REGIONS OF THE DOG

As a slightly different approach to that adopted in fig 1, in which a selection of 'points' was indicated, these four illustrations show the body surface divided up into a number of regions which are given particular names. Nevertheless, as you may notice, many of the points of the dog are represented in this drawing as regions of the body because of their distinctness from neighbouring areas.

Some of the names used for body regions refer specifically to the major internal structures found in that particular region; eg. the femoral region based on the femur (thigh bone) and its surrounding muscles; the parotid region underlain by the parotid salivary gland, and so on. For describing positions on the body surface, and for describing the position of internal organs in relation to the surface, these named regions often prove very useful.

Skin contains a selection of **glands** situated in the dermis and emptying their secretions onto the epidermal surface through pores. *Sebaceous glands* are present throughout the skin and open into the hair follicles. The fatty semi-liquid secretion, *sebum*, solidifies when exposed to air and becomes applied to the hair root and the neighbouring epidermal surface. It helps to keep the skin soft and pliable and to 'waterproof' the body. Spread thinly over individual hairs sebum prevents them from becoming brittle and is responsible for the glossy sheen of the coat. The *tarsal (Meibomian) glands* along the internal edge of the eyelids are modified and specialized sebaceous glands producing an oily secretion. The oily superficial layer this imparts to the tears film of the eye reduces evaporation, lubricates the lids preventing them from sticking, and forms a barrier at the lid margins preventing tear overflow onto the face.

Sweating is a method used by many animals to lower their body temperature. The evaporation of fluid (sweat) from the skin gives a considerable cooling effect. However, in dogs true *sweat glands* are only found in the pads on the paws where their watery secretion may play a role in gripping. As an alternative method of cooling dogs pant, the evaporation of water from the epithelial surfaces of the lips, tongue, mouth cavity and lungs, accomplishes a similar cooling effect to sweating. Nevertheless dogs are prone to overheating.

Several further types of gland are modifications of sweat glands. *Odoriferous* glands are quite widespread and open into hair follicles. The scent derived from their secretions is of importance in social communication between dogs. Of more restricted occurrence are the *ceruminous glands* of the ear canal producing the ear wax (*cerumen*), and the *glands of the anal sacs (paranal sinuses)*. The latter produce the foul-smelling secretion which collects in the paired anal sacs on either side of the anal canal. A duct from each anal sac opens onto the skin each side of the anus and the secretion is added to the surface of faeces as they are voided.

Anal sacs are distinct from *circumanal glands* opening separately onto the skin at and around the anus. These are modified sebaceous glands (odoriferous glands) which produce a particularly attractive secretion — at least to other dogs! Finally the *mammary glands* have a very similar structure to sweat glands and develop from the same rudiments. Five pairs of glands are a normal complement (fig 17D), although four pairs or even six pairs are not unusual, lying in the superficial fascia beneath the dermis. A conical *teat* marks the position of each gland but even these are only distinct projections in a lactating bitch and are rudimentary in a dog.

Topographical regions of the head
1–5 Cranium. **1** Frontal (supraorbital region). **2** Parietal region. **3** Occipital region. **4** Temporal region. **5** Auricular region. **6–21** Face. **6–8** Nasal region. **6** Dorsal nasal region. **7** Lateral nasal region. **8** Nostril region. **9–10** Oral region. **9** Upper lip. **10** Lower lip. **11** Mental region. **12–13** Orbital region. **12** Upper eyelid. **13** Lower eyelid. **14** Zygomatic region. **15** Infraorbital region. **16** Temporomandibular (jaw) joint. **17** Masseteric region. **18** Buccal region. **19** Maxillary region. **20** Mandibular region. **21** Intermandibular region.

Topographical regions of the neck
22 Dorsal neck region. **23** Lateral neck (jugular) region. **24** Parotid region. **25** Pharyngeal region. **26–27** Ventral neck region. **26** Laryngeal region. **27** Tracheal region.

Topographical regions of the thorax (pectoral regions)
28 Presternal region. **29** Sternal region. **30** Scapular region. **31** Costal region. **32** Cardiac region.

Topographical regions of the abdomen
33–34 Cranial abdominal (epigastric) region. **33** Hypochondriac region. **34** Xiphoid region. **35–36** Middle abdominal (mesogastric) region. **35** Lateral abdominal (iliac) region (includes paralumbar fossa). **36** Umbilical region. **37–39** Caudal abdominal (hypogastric) region. **37** Inguinal region. **38** Pubic region. **39** Preputial region.

Topographical regions of the back (dorsal regions)
40 Interscapular region. **41** Thoracic vertebral region. **42** Lumbar region.

Topographical regions of the pelvis and tail
43 Sacral region. **44** Gluteal region. **45** Coxal tuberosity region. **46** Clunial region including ischiorectal fossa. **47** Ischiatic tuberosity region. **48–50** Perineal region. **48** Anal region. **49** Urogenital region. **50** Scrotal region. **51** Caudal region.

Topographical regions of the forelimb (thoracic limb)
52 Shoulder joint. **53** Axillary region (includes axillary fossa). **54** Brachial region. **55** Tricipital region. **56** Cubital region. **57** Olecranon region. **58** Antebrachial region. **59** Carpal region. **60** Metacarpal region. **61** Phalangeal region (digits).

Topographical regions of the hindlimb (pelvic limb)
62 Hip joint. **63** Femoral region. **64** Genual region. **65** Popliteal region. **66** Patellar region. **67** Crural region. **68** Tarsal region. **69** Calcaneal region. **70** Metatarsal region. **71** Phalangeal region.

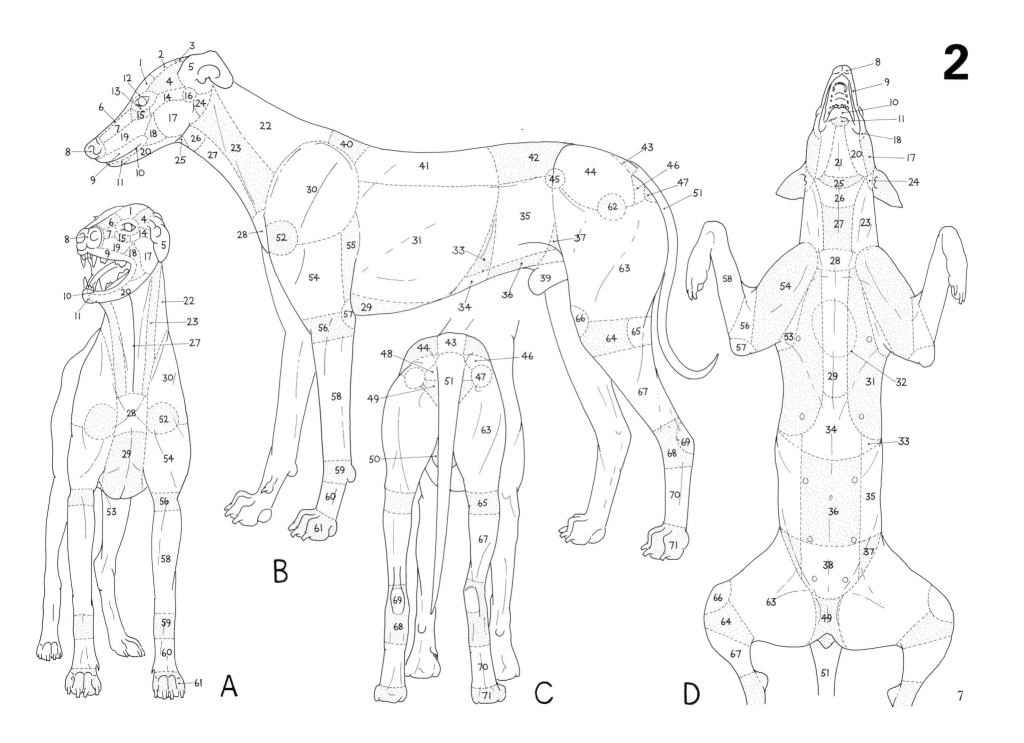

3

SKELETON OF THE DOG

The skeleton is a scaffolding of hard structures upon which the soft tissues are hung. It therefore provides a supporting framework for the body, sites for the attachment of muscles, as well as protecting some of the internal organs. **Bone** is the major structural material making up the skeleton although cartilage plays an important role in both foetal (unborn) and young animals, and also in certain parts of an adult skeleton. If you remember, bone was considered in the introduction to be a type of connective tissue, but one in which the material surrounding cells is heavily impregnated with calcium salts. Thus, despite bone being a living tissue with a permeating network of blood vessels and nerves, considerably more than half its weight is non-living matter, mainly a complex crystalline compound of calcium phosphate produced by special bone-forming cells. From a mechanical standpoint this inorganic component gives bone the degree of rigidity necessary for it to withstand considerable compression forces. The organic or living component of bone, composed of cells, fibrous material and intercellular fluid, gives bone a certain degree of flexibility preventing it from being brittle.

The entire skeleton is shown in this illustration from the left side as if in a normal standing position. If you compare this drawing with fig 1 you will see that a great many bones, and especially their irregularities of contour, may be felt through the dog's skin. It might be a useful exercise on your part if you were to take a pencil and shade in on the drawing the areas of the skeleton that you can definitely feel through the skin of your own animal. You might also see that according to its position in the body the skeleton is made up from two components: an **axial skeleton** forming the skeletal basis of the head, neck, trunk and tail, and consisting of the skull, vertebral column, ribs and sternum; and an **appendicular skeleton** comprising the bones of the limbs and associated girdles attaching the limbs to the trunk.

An illustration such as this is comparable in many ways to the mounted skeleton of a dog that you might see displayed in a museum. As I shall explain later, it is a somewhat artificial picture in that skeletal bones are held together in life by ligaments and muscles, neither of which are shown in this drawing although later drawings will demonstrate both. In a museum preparation of a skeleton the role of ligaments and muscles is undertaken by glue and wire, the preparator attempting to connect the bones together in as normal an arrangement as possible.

In order to perform their purely mechanical functions of support and muscle attachment most bones do not require to be solid masses of bone tissue. In fact this would be a drawback since it might well add substantially to overall body weight, whilst adding little to the strength of a bone. Therefore bone tissue may vary somewhat in structure and disposition depending upon the stresses and strains imposed on it during life. Areas of bone which are heavily stressed such as the outer shell of all skeletal bones, are formed of very hard, dense bone tissue called **compact bone**. Further increase in thickness of this peripheral layer at particular points, the noticeable lumps and bumps of a bone, denote areas which are subjected to even greater stress such as that produced by the pull from muscle tendons or from ligament attachment. Inside the compact outer layer all bones have a more loosely arranged network of **spongy bone** tissue with numerous honeycomb-like spaces. This more randomly arranged type of tissue forms in areas which are receiving less stress. Some areas in the interior of bones may be completely unstressed and here bone tissue is absent altogether leaving larger spaces. These internal cavities are available for housing other tissues, especially fragile and easily damaged ones. Thus blood forming tissue (*red bone marrow*) is contained within *marrow spaces* inside the long bones of limbs, the ribs, sternum and skull roofing bones. As a dog ages, blood forming tissue tends to become more restricted in its distribution so that cavities inside bones may become vacated and available for the storage of fat (*yellow bone marrow*). This change from red to yellow marrow happens particularly in the long bones of the limbs, although red marrow persists inside ribs, sternum and vertebral bones throughout life.

Although there are these variations in bone as a tissue, the drawings show bones as organs in their own right, each one made up predominantly from bone tissue, but also possessing blood vessels, nerves, bone marrow, and an outer limiting membrane, the **periosteum**. This tough investing layer of fibrous connective tissue covers the entire surface of a bone, except where it forms a joint, and is intimately attached to it, periosteal fibres penetrating into the compact peripheral bone tissue itself. Where a tendon or ligament attaches to a bone, penetrating periosteal fibres are most numerous and appear to form a direct continuation of the tendon or ligament into the bone substance – an arrangement ensuring the necessary firmness of attachment for muscles onto bones. The innermost periosteal layer, although very delicate, is extremely important since it contains cells able to produce new bone. The limiting function of the outer fibrous layer is therefore necessary to keep such bone-forming cells confined within it. Should such tissue 'escape', as may occur if periosteal rupture accompanies a bone fracture, then extra bone may be produced outside the periosteum. The periosteum is also extremely sensitive: bone tissue itself is insensitive, sensation from bone arising only from nerve endings in its periosteum. Hence, unless this outer membrane is involved, disease in bone could well be painless.

It is also apparent from the drawing that skeletal bones assume a variety of different shapes and sizes and present a bewildering array of lumps and bumps on their surfaces. **Long bones** are the elongated, cylindrical limb bones such as the humerus or femur, basically consisting of an outer compact layer surrounding an internal marrow cavity. They form strong rods resisting compression (squashing) stress along their long axes when supporting body weight, and resisting tension (bending or stretching) stress at angles to their long axes when muscles attaching to them contract. At their ends long bones are generally enlarged: a smooth surfaced expansion indicates an articulatory area with an adjoining bone, whilst a more roughened enlargement provides an area for tendon or ligament attachment. The expanded ends contain spongy bone which, because of its lightness, allows the extremity to be enlarged to increase the surface area of a joint, improving joint stability without adding unduly to overall bone weight.

Short bones are approximately equal in all dimensions and are found only in the carpus (wrist) and tarsus (ankle) regions. Here they seem to play a role in reducing or dissipating concussive forces, and the consequent possibility of injury to a leg as shock waves pass up its skeleton when the paw hits the ground. A special type of short bone found near freely moving limb joints is a **sesamoid bone**. These, like the patella ('knee cap'), are usually located in a tendon although they may develop in the ligamentous tissue over which a tendon passes. Sesamoids protect tendons where they are subjected to friction, as in their passage over bony prominences, since bone is a tissue with the capacity to withstand friction and pressure to a much greater degree than can fibrous tendon.

Flat bones are found in the limb girdles and in the head and thorax. The ribs and bones of the face and roof of the cranium are obviously protective, but they also provide an increased surface area for muscle attachment. In the cranium the flat bones consist of inner and outer layers of compact bone sandwiching a spongy layer. In places the spongy bone is invaded by air-filled extensions from the nasal cavity widely separating the compact layers. Air spaces inside bones are termed sinuses and those associated with the nasal region are **paranasal sinuses**.

A final category called **irregular bones** includes those unpaired midline bones of the vertebral column and base of the skull characterised by numerous jutting processes for muscle and ligament attachment as well as for articulation.

Although the drawing does not make a distinction between cartilage and bone we have already suggested that cartilage forms an important skeletal component. In terms of its mechanical properties it occupies an intermediate position between dense fibrous connective and bone, being able to withstand considerable amounts of compression and bending like bone, whilst retaining a considerable degree of flexibility like fibrous connective tissue. It is abundant in a developing animal, many parts of the skeleton developing initially as a preliminary cartilaginous framework. As a foetus grows in size its cartilage is gradually replaced by bone, the cartilage having formed a scaffolding upon which bone is constructed. Some parts, however, remain cartilaginous throughout life and of special importance in this respect is that it contributes an important functional component to many joints. A possible explanation for its limited occurrence in an adult might be that, unlike bone, it does not have a network of capillary blood vessels running through it. Blood capillaries terminate at its periphery so that nutrients can only pass to embedded cartilage cells, and waste products pass out from them, by passage through the fluid component of its intercellular material. However, to perform its mechanical functions this fluid matrix of cartilage is thickened considerably to a glue-like consistency. The diffusion pathway through this thick, semifluid material to and from blood vessels at the cartilage surface can only be short, thus restricting the depth that cartilage can attain. Should it expand beyond this size then cells in its interior will be cut off from their nutrient source and will die. Cartilage cell death is in some way related to the deposition of calcium salts and so **calcification** is a common occurrence in large masses of cartilage especially as a dog ages. This relationship between cartilage cell death and calcium deposition is also important in the normal replacement of cartilage by bone during development.

The skeleton of a dog sometime before birth consists of cartilage and connective tissue only. At some stage during development in the uterus special cells initiate the process of bone formation – **ossification**. Bone therefore forms in connective tissue or cartilage, both processes being essentially similar, although in the latter the cartilage must first be removed before bone can be formed. In either case bone formation begins at specific centres of ossification, each definitive bone having one major centre to start with, and one or more subsidiary centres which form and join with it during growth.

In connective tissue, bone forming cells begin the mineralization of surrounding tissue producing a calcified area around themselves. Once surrounded by rigid matrix the cells can no longer divide but remain alive and in connection with neighbouring cells and with capillary blood vessels occupying minute tunnels in the developing tissue. Subsequent growth of this centre of ossification can only occur on its outer surface by the plastering of new bone onto that already formed. An area of bone soon forms separated from adjoining areas by persistent connective tissue. The superficial layer at the surface of an expanding area of new bone forms a periosteum and bone production continues from its inner layer until the bone reaches its adult size. Bones such as those of the face form by this method of ossification in connective tissue.

In those areas of cartilage in a foetal skeleton the first stage in bone formation must be cartilage removal to make way for a centre of ossification. Subsequently, bone forming cells begin to produce bone as they do in connective tissue. As this primary bone centre inside an area of cartilage increases in size by surface deposition, the cartilage is gradually eroded to make way for its expansion. But, if cartilage does not itself continue to grow it will be replaced by bone in a very short time. It is essential therefore that cartilage should itself proliferate at least as fast as it is being destroyed and replaced by bone. At some stage in development secondary centres of ossification appear in the ends of bones where they are forming joints or attaching tendons or ligaments. A developing bone now has three clear parts: a central **diaphysis** and two extremities (**epiphyses**). Continued growth in length is ensured by the maintenance of an actively growing plate of cartilage between epiphysis and diaphysis to provide cartilage for subsequent conversion into bone. As maturity is reached the rate of bone deposition ultimately overtakes that of cartilage cell multiplication and growth, and ossification completely invades the growth plate of cartilage. Bone growth in length effectively ceases, but growth in bone diameter can continue to occur by deposition of bony tissue at the surface beneath the periosteum.

Bone once formed is not an unchanging structure, it undergoes structural rearrangement to combat the differing stresses encountered during growth. Provision is made for this change in the form of bone destroying cells sited in the periosteum. These cells can erode existing bone, dissolving its rigid matrix and making way for new bone forming cells to lay down bone in different patterns or even in different places. Thus the interior of the shaft of a long bone, for instance, receives very little stressing as it enlarges, so that bone initially deposited here is reabsorbed as the bone increases in girth by deposition at its outer, periosteal surface. This reabsorption produces a cavity at the centre of a bone which is available for housing bone marrow as we noticed earlier.

A bone's capacity to change in form according to an animal's needs is important should it be fractured. Bone in the region of the break is literally 'dissolved' and completely rebuilt to unite the broken parts. Over a period of several months the new bone is stressed and gradually remodelled to conform to the original shape of the bone. If reset in the correct alignment following the initial fracture, the final result may well be practically indistinguishable from uninjured bone.

Skull
1 Cranium (braincase). **2** Occiput (caudal boundary of cranium). **3** Tympanic bulla (surrounding middle ear [tympanic] cavity containing 3 auditory ossicles – incus, malleus and stapes). **4** Face (muzzle – based on nasal cavity and jaws and attached to rostral end of cranium). **5** Nasal cartilages (movably articulated with incisive bones of face and surrounding nasal vestibule). **6** Nasal cavity (containing olfactory apparatus and forming initial part of respiratory air tract). **7** Orbit (housing and protecting eyeball – continuous caudally with temporal fossa). **8** Upper jaw (supporting upper dental arch – consisting of 6 incisor teeth, 2 canine teeth, 8 premolar and 4 molar teeth). **9** Mandible (lower jaw supporting lower dental arch – consisting of 6 incisor teeth, 2 canine teeth, 8 premolar and 6 molar teeth). **10** Hyoid apparatus (suspending tongue and larynx in floor of throat). **11** Thyroid cartilage (most prominent cartilage of larynx – 'voice box').

Vertebral column, ribs and sternum
12–15 Cervical (neck) vertebrae. **12** Atlas (cervical vertebra 1). **13** Axis (cervical vertebra 2). **14** Cervical vertebra 4. **15** Last (7th) cervical vertebra. **16–18** Thoracic (dorsal or back) vertebrae. **16** Summit of spinous process of thoracic vertebra 1. **17** Anticlinal vertebra (Thoracic vertebra 10). **18** Last (13th) thoracic vertebra. **19–20** Lumbar (loin) vertebrae. **19** Lumbar vertebra 2. **20** Summit of spinous process of last (7th) lumbar vertebra. **21** Sacrum (3 fused sacral vertebrae in pelvic region). **22–24** Caudal (tail) vertebrae. **22** Caudal vertebra 1. **23** Caudal vertebra 6. **24** Caudal vertebra 18. **25–30** Thoracic rib-cage formed from 13 pairs of ribs. **25** Rib 1. **26** Bony part of rib 6. **27** Costal cartilage at lower end of rib 6. **28** Costal cartilage of rib 9 (last sternal [true] rib – ie. with direct attachment to sternum). **29** Costal arch (formed from fusion of costal cartilages of ribs 10–12, asternal [false] ribs – ie. without direct attachment to sternum, only indirectly through association with costal cartilage of rib 9). **30** Rib 13 (last or floating rib connected by fibrous tissue with costal arch). **31–33** Sternum ('breastbone' formed from 8 individual sternal segments [sternebrae] joined by intersternebral cartilages). **31** Manubrium of sternum (sternebra 1 elongated into base of neck). **32** Sternebra 3. **33** Xiphoid cartilage of sternum (cartilaginous prolongation into belly wall of 8th [last] sternebra, xiphoid process).

Joints of axial skeleton
34 Jaw (temporomandibular) joint. **35** Atlantooccipital joint ('yes' joint). **36** Atlantoaxial joint ('no' joint). **37** Costovertebral joint. **38** Costochondral joint. **39** Sternocostal joint.

Forelimb skeleton
40 Scapula (shoulder blade of pectoral girdle). **41** Humerus (arm bone). **42** Radius. **43** Ulna. **44–47** Carpus ('wrist' – based on 7 carpal bones arranged in two rows). **44** Radiocarpal bone. **45** Ulnar carpal bone. **46** Accessory carpal bone. **47** Carpal bones 1–4. **48–49** Metacarpus ('palm' – based on 5 metacarpal bones). **48** Metacarpal bone 1. **49** Metacarpal bone 5. **50–53** Phalanges (3 in each digit except digit 1 ['dewclaw'] with only 2). **50** Proximal (1st) phalanx of digit 5. **51** Middle (2nd) phalanx of digit 5. **52** Distal (3rd) phalanx of digit 5. **53** Proximal palmar sesamoid bones of digit 5 (a pair at each metacarpophalangeal joint in tendon of insertion of interosseous muscle [proximal plantar sesamoids at equivalent positions in hindpaw]).

Joints of forelimb
54 Shoulder (scapulohumeral) joint. **55–57** Elbow joint (composite joint with 3 interrelated components). **55** Humeroulnar joint. **56** Humeroradial joint. **57** Proximal radioulnar joint. **58** Distal radioulnar joint. **59** Antebrachiocarpal joint (dominant component of composite carpal joint). **60** Metacarpophalangeal joint, digit 2. **61** Proximal interphalangeal joint, digit 2. **62** Distal interphalangeal joint, digit 2.

Hindlimb skeleton
63–65 Hip bone (pelvic bone of pelvic girdle – formed from 3 separate bones which fuse together during development). **63** Ilium. **64** Ischium. **65** Pubis. **66** Femur (thigh bone). **67** Patella ('knee cap' – sesamoid bone in tendon of insertion of quadriceps femoris muscle). **68** Fabellae (stifle sesamoid bones in tendons of origin of gastrocnemius muscle). **69** Tibia. **70** Fibula. **71–74** Tarsus (ankle or hock – based on 7 tarsal bones arranged in 3 rows). **71** Talus (astragalus or tibial tarsal bone). **72** Calcaneus (os calcis or fibular tarsal bone). **73** Central tarsal bone. **74** Tarsal bones 1–4. **75–77** Metatarsus ('sole' – based on 5 metatarsal bones). **75** Metatarsal bone 1 (rudimentary). **76** Metatarsal bone 2. **77** Metatarsal bone 5.

Joints of hindlimb
78 Hip (coxofemoral) joint. **79–81** Stifle (knee) joint (composite joint with 3 components). **79** Femorotibial joint. **80** Femoropatellar joint. **81** Proximal tibiofibular joint. **82** Distal tibiofibular joint. **83** Talocrural joint (dominant component of composite tarsal joint). **84** Metatarsophalangeal joint of digit 5.

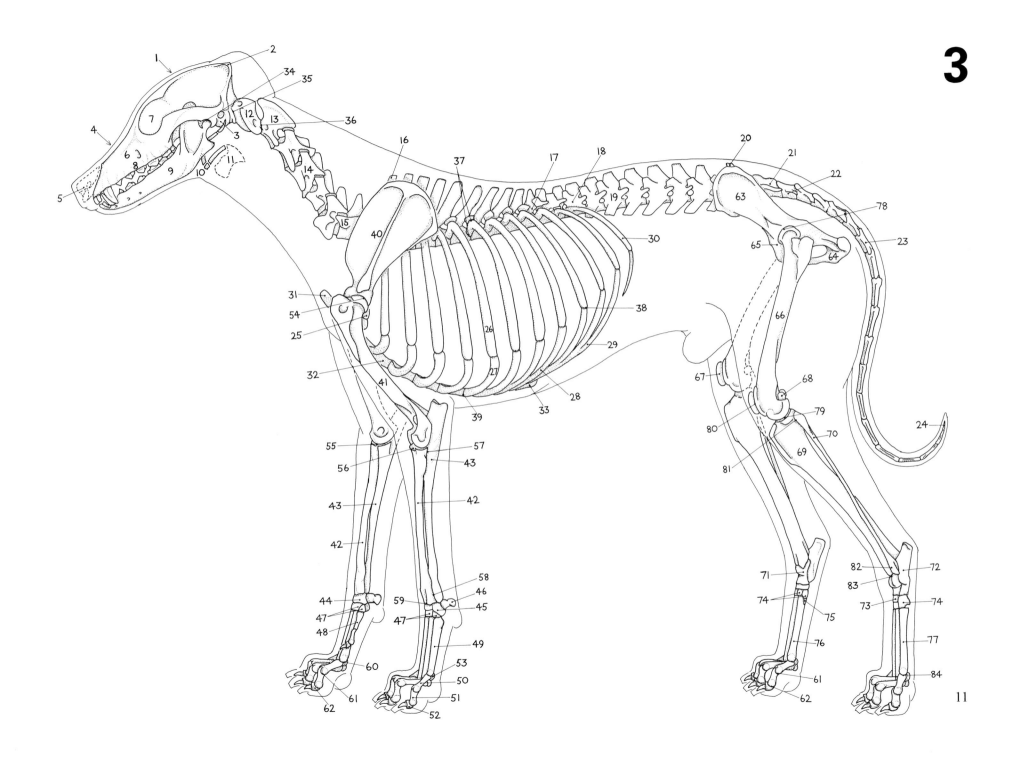

3

11

4

AXIAL SKELETON OF THE DOG – SKULL, VERTEBRAL COLUMN, RIBS AND STERNUM

The main illustration here (A) shows the vertebral column (spine or backbone), ribs and sternum from the side, the limbs and girdles (appendicular skeleton) having been removed. The accompanying drawings at the foot of the page (D & E) show the first two neck vertebrae, the axis and atlas in enlarged view, and a surface view of the trunk and tail on which are indicated in solid black those 'landmark' features of the vertebral column and ribcage that are palpable through the skin. The skull, a structure of some complexity, is drawn in lateral view (B) together with the nasal and laryngeal cartilages in their appropriate positions (the auricular cartilage has been removed). Only the major points of the skull are included in the legend; individual bones are not separately labelled since, in the adult skull, the individual bones have fused together to give a single structure. Consequently an additional surface view of the head (C) is given on which the strategic 'landmark' components that are palpable through the skin are inked in. Obviously these are only the major landmarks, other expanses of bone that are palpable are indicated by the stippled areas. I'm sure that with very careful palpation of your own animal you could identify more of the skull. For instance it may be possible to palpate practically all of the hyoid bones, much of the facial bones, and some of the bones of the cranium.

The **vertebral column** is clearly made up of a chain of **vertebrae**, unpaired irregular bones extending from the back of the skull to the tip of the tail. The 'artificial' nature of skeletal drawings is again apparent since in life these bones would only be held in this position by numerous ligaments and muscles. Changes in position will depend upon these ligaments and muscles being able to act as adjustable ties. Despite a number of differences, the structure of all vertebral bones is essentially based on a common plan related to certain fundamental functions performed by the spine.

Firstly, the backbone provides the main longitudinal girder of the body, strong enough to support weight suspended from it and sufficiently rigid so that propulsive forces from the limbs will not buckle it. Each vertebra therefore has a solid, cylindrical **vertebral body** joined to adjacent bodies in front and behind by intervertebral discs of fibrocartilage. As well as anchoring vertebral bodies together **intervertebral discs** also allow limited movements between vertebrae. The amount of movement that any single disc permits is obviously very slight. Nevertheless the sum total of all of these small movements at individual discs amount to the considerable flexibility of the whole spine. Thus the back can flex and extend (arching and straightening) and also bend from side to side, although twisting (rotation) between vertebrae is a movement which is severely restricted in extent as we shall see later.

Secondly, each vertebral body is surmounted dorsally by a **vertebral arch** which surrounds the spinal cord in a canal. The arches of adjacent vertebrae form a continuous tunnel along the length of the back housing and protecting the spinal cord. Since nerves and blood vessels must be able to pass into and out of the vertebral canal to connect with the spinal cord, the arches cannot be complete. Each one has cranial and caudal notches on either side close to the attachment of arch to body. The notches of adjacent vertebrae surround an **intervertebral opening** leading into the vertebral canal.

Thirdly, the backbone provides attachment for muscles and ligaments on both vertebral body and arch, and also where the arch is drawn out dorsally as a **spinous process** and on either side as **transverse processes**. Vertebral ligaments include three longitudinal components: *dorsal* and *ventral longitudinal ligaments* joining vertebral bodies and the *supraspinous ligament* linking the tips of spines. The transverse processes on certain vertebrae additionally provide attachment for **ribs** which are bones transferring muscle forces to the backbone.

Finally, the backbone must also resist any tendency for one vertebra to twist in relation to its neighbours, a movement which would massively disrupt intervertebral discs and would severely damage blood vessels and nerves passing through intervertebral openings. **Articular processes** are specifically designed to restrict rotation since they are overlapping projections of bone; a pair of cranial processes of one vertebra overlap a pair of caudal processes of the vertebra in front. Although restricting twisting they still allow limited vertical and/or horizontal movements between vertebrae.

The backbone is generally considered as having five distinct regions corresponding to the areas in the body in which they are found, and the number of vertebrae in each region is constant:
Cervical (neck) region with 7 vertebrae.
Thoracic (back or chest) region with 13 vertebrae.
Lumbar (loin) region with 7 vertebrae.
Sacral (pelvis or croup) region with 3 fused vertebrae.
Caudal (tail) region with approximately 20 vertebrae.
The reasons underlying spinal regionalization are based upon different functions which predominate in the various body areas.

Independent head movement requires a well developed series of neck bones extending forwards from the trunk. Head mobility has also been increased by reducing the length of spinous processes dorsally, by a complete loss of free ribs laterally, and by specialization of the first two neck vertebrae, the **atlas** and **axis**. These two vertebrae are shown in **D**, a drawing in which the left half of the atlas has been removed opening the spinal canal and displaying the two important joints in terms of head mobility. The **atlantooccipital joint**, between enlarged cranial articular processes of the atlas and occipital condyles of the skull, allows movement of the head in a vertical plane (nodding) – the 'yes' joint: the **atlantoaxial joint**, between the odontoid process of the axis and the ventral arch of the atlas, allows rotation in a horizontal plane (shaking the head) – the 'no' joint. At first sight this latter joint would seem to contradict the statement made earlier that rotation between vertebrae is not permissible. However, this is in fact the *only* intervertebral joint where it is permitted and an intervertebral disc is absent. As you may also see from the drawing the intervertebral opening between atlas and axis is an extremely large one in which nerve and vessel damage will be avoided.

The reduction of vertebral spines together with the need to support the weight of the head on the end of a mobile neck has also led to a modification of the supraspinous ligament in the neck. It now extends forwards from thoracic spines as a **nuchal ligament** attaching only to the spine of the axis.

The chest, based on **ribs**, is an area concerned with breathing movements. Lungs are contained within airtight cavities inside the chest so that on thoracic enlargement they expand, their internal pressure is lowered, and air is drawn into them. The structural requirement for such a mechanism is an encircling box resistant to collapse while at the same time being able to increase and decrease in size. Ribs are therefore well-formed and encircle the chest. They are arranged in 13 pairs, corresponding in number to thoracic vertebrae, and each articulates above with a transverse process and a vertebral body. Below, the first nine pairs of ribs contact the **sternum** (*breastbone*) directly, the *sternal* or *true ribs*, whilst the lower ends of the next three are linked as a *costal arch* which only indirectly contacts the sternum through the articulation with the costal

cartilage of rib 9, the *asternal (false) ribs*. The last (*floating*) rib usually ends freely in the musculature of the flank although it has a fibrous connection with the costal arch.

Ribs and sternum constitute the **ribcage** and, together with thoracic vertebrae, form the bony boundaries of a thoracic cavity of which we will have much more to say later on. The **sternum** is a midline structure made up of 8 pieces or *sternebrae* connected by cartilage. The first sternebra is expanded in front as the *manubrium* into the base of the neck; the last is expanded behind as a *xiphoid process* into the abdominal wall. Although separate in a young animal, individual sternebrae tend to fuse together so that an aged dog's sternum may appear to be a single structure.

A second important function undertaken by the ribcage is that of transferring body weight onto the frontlegs. The trunk is supported on the frontlegs by **ventral serrate muscles**, one on either side, running upwards from bony ribs to shoulder blades. Trunk weight is transferred from bony ribs through these muscles to the upper ends of the limbs. The upper bony part of a rib is therefore under quite considerable compression (squashing) stress and we have already suggested that bone is the most suitable material for withstanding this. Below the attachment of the ventral serrate muscle ribs are subjected to much less stress and remain cartilaginous as the **costal cartilages**.

Efficient support of body weight and its transfer to the legs is of primary importance as has just been mentioned. Much of the trunk weight is concentrated in the abdominal viscera. How-

ever, since these organs are mobile and undergo frequent changes in dimension they are supported by a muscular framework rather than a more rigid and restricting framework of ribs. The muscular abdominal wall is slung from the prominent transverse processes of enlarged lumbar vertebrae. Throughout both thoracic and lumbar regions large masses of muscle lie above the vertebrae and the processes present for their attachment are large and well-formed.

Propulsive thrust generated by the hindlegs is transmitted to the backbone in the trunk. This power transfer is made more effective by firmly anchoring the pelvic girdle to the sacral spinal region. Attachment is through enlarged transverse processes, *sacral wings*, with which the ilia of the pelvic bones form **sacroiliac joints**. A greater area of anchorage, thus improving stability, is given by incorporating three sacral vertebrae into a compound **sacrum**.

As with the neck, mobility is a primary requirement of the tail which means that all projecting processes are reduced simplifying vertebral structure. But, unlike the neck, little if any weight has to be carried by the tail so that caudal vertebrae become simplified until towards the tip of the tail they are only represented by elongated vertebral bodies.

Overall the backbone can be likened to a longitudinal beam composed of a connected chain of bones capable of withstanding compression – it wont concertina when pushed. This is particularly important during locomotion since it transmits the thrust delivered by the hindlimbs to the trunk. From

this beam running along the axis of the body, strut-like processes project upwards or outwards. In this mechanical analogy ligaments and muscles of the back act as ties connecting the struts enabling the spine to resist tension – it can control bending forces. Between them they can effectively immobilize the backbone so that it may support body weight in its postural capacity. Of special importance in maintaining the curvatures of the spinal column are a number of spinal ligaments and the cooperative action between epaxial muscles dorsal to the backbone, subvertebral components of hypaxial muscles immediately beneath the backbone, and abdominal muscles particularly the straight (rectus) muscles. Variations in the arch of the spine will be effected by changes in tension and action within these three groups of muscles adding to spinal mobility during movement.

The vertebral column is not a straight structure but has curves in it. In a normal standing position a dog holds its head up, the cervical part of its spine is consequently at an angle to its thorax, a curvature occurring at the base of the neck. In order for the head to face forwards there is also a slight curvature at the head end of the neck. At the hind end of the body the tail is also somewhat similarly S-shaped in many breeds, the tip of the tail pointing upwards. Between neck and tail the spine in the trunk is slightly curved with the concavity below. Although a single smooth curve, it seems to be made up of two functional parts more or less corresponding to the two structural units, lumbar and thoracic. The thoracic unit projects back as a girder-like support from an anchorage point at the upper ends of the forelimbs, while a lumbar bracket projects forwards from an anchorage at the sacrum above the hindlimbs. The size and strength of the backbone at any point along its length is clearly related to the stresses and strains that it bears; the greatest stress no doubt occurring at the points where it attaches to the limbs. Thus in the loins each vertebral bone carries the weight of the body in front of it together with the weight of its own body 'segment'. Passing back towards the sacrum each vertebra is accepting more total weight than the one in front of it and you can see from the drawings that the vertebrae enlarge back through the lumbar region. Likewise in the thorax, the vertebrae, and in particular their spines, are at their largest and longest over the shoulders where strains are greatest.

Skull

1–15 Face (based on nasal cavity and jaws). 1 Nasal cartilages (movably articulated with bone of nasal cavity and surrounding nasal vestibule). 2 Nasal process of incisive bone (bordering bony nasal opening leading into nasal cavity proper). 3 Alveolar border of maxillary bone (upper jaw bearing teeth of upper dental arch). 4 Upper carnassial (shearing) tooth (premolar 4). 5 Alveolar jugae (impressions on surface of maxillary bone created by tooth roots internally). 6 Infraorbital foramen. 7–13 Mandible (lower jaw). 7 Body of mandible. 8–11 Ramus of mandible. 8 Angular process of mandible. 9 Coronoid process of mandible (insertion of temporal muscle). 10 Masseteric fossa of mandible (insertion of masseter muscle). 11 Jaw (temporomandibular) joint (condyloid process of mandible located in mandibular fossa of squamous temporal bone). 12 Mandibular symphysis (fibrocartilaginous intermandibular joint allowing practically no movement). 13 Alveolar border of mandible (bearing teeth of lower dental arch). 14 Lower carnassial (shearing) tooth (molar 1). 15 Mental foramen. 16–30 Cranium (braincase). 16 Zygomatic arch (bridge of bone connecting face and cranium below eye). 17 Orbit (housing and protecting eyeball). 18 Zygomatic (supraorbital) process of frontal bone. 19 Temporal line (rostral divergence of external sagittal crest). 20 External sagittal crest (in dorsal midline of cranium). 21 External occipital protuberance (occiput – most dorsocaudal portion of cranium). 22 Nuchal crest (division between dorsal and caudal surface of cranium). 23 External occipital crest. 24 Temporal fossa (origin of temporal muscle). 25 Jugular process of occipital bone. 26 Occipital condyle (both condyles form atlantooccipital joint with atlas). 27 Foramen magnum. 28 Mastoid process of temporal bone (sole representation on skull surface of petrous temporal bone). 29 Tympanic bulla (surrounding tympanic or middle ear cavity and containing 3 auditory ossicles – incus, malleus and stapes). 30 External acoustic meatus (across which eardrum is stretched in life and around which auricular cartilage is attached). 31–36 Hyoid apparatus supporting tongue and larynx. 31–34 Cranial horn of hyoid lying in pharyngeal wall. 31 Ceratohyoid bone. 32 Epihyoid bone. 33 Stylohyoid bone. 34 Tympanohyoid cartilage (articulating with tympanic bulla and mastoid process). 35 Basihyoid bone (lying transversely in floor of throat). 36 Caudal horn of hyoid (thyrohyoid bone in wall of laryngopharynx articulating with thyroid cartilage of larynx). 37–39 Laryngeal cartilages (forming larynxvoicebox). 37 Epiglottic cartilage. 38 Thyroid cartilage. 39 Cricoid cartilage.

Vertebral column

40–52 Cervical (neck) vertebrae. 40 Dorsal arch of atlas (cervical vertebra 1 – cut through in median plane in fig D). 41 Dorsal tubercle of atlas (spinous process absent). 42 Wing of atlas (enlarged flattened transverse process). 43 Ventral arch of atlas (vertebral body absent – cut through in median plane in fig D). 44 Alar notch of atlas. 45 Spinous process of axis (cervical vertebra 2). 46 Odontoid process (dens) of axis (developmentally represents vertebral body of atlas). 47 Cranial articular surface of axis. 48 Vertebral canal in atlas. 49 Spinous process of last (7th) cervical vertebra (much reduced). 50 Transverse process of cervical vertebra 6 (enlarged and bifid). 51 Transverse foramen of cervical vertebra (consecutive foramina producing a transverse or vertebrarterial canal for passage of vertebral artery and vein). 52 Transverse (vertebrarterial) canal of axis vertebra. 53–56 Thoracic (chest or back) vertebrae. 53 Spinous process of thoracic vertebra 1. 54 Spinous process of thoracic vertebra 10 (anticlinal vertebra). 55 Spinous process of last (13th) thoracic vertebra. 56 Transverse process of thoracic vertebra 5. 57–58 Lumbar (loin) vertebrae. 57 Spinous process of last (7th) lumbar vertebra. 58 Transverse process of lumbar vertebra 5. 59–62 Sacrum (3 fused sacral vertebrae). 59 Median sacral crest (3 fused spinous processes of sacral vertebrae). 60 Sacral wing (enlarged 1st sacral transverse process). 61 Lateral sacral crest (fused sacral transverse processes 2 and 3). 62 Auricular surface of sacral wing (for formation of sacroiliac joint). 63 Caudal (tail) vertebrae. 64–67 Additional vertebral processes. 64 Cranial articular process of vertebra. 65 Caudal articular process of vertebra. 66 Accessory process of vertebra (present only on caudal thoracic and lumbar vertebrae). 67 Mamillary process of vertebra (present on thoracic and lumbar vertebrae). 68–70 Intervertebral foramina (for passage of spinal nerves, arteries and veins). 68 Intervertebral foramen 1 (lateral vertebral foramen of atlas). 69 Intervertebral foramen 2 (between atlas and axis – for passage of cervical nerve 2). 70 Intervertebral foramina 8 (last cervical), 17 (9th thoracic) and 25 (4th lumbar).

Ribs and sternum

71 Rib head (capitulum) of rib 5. 72 Rib tubercle of rib 5. 73 Shaft (bony body) of rib 1. 74 Costal cartilage of rib 6 (articulating with intersternebral cartilage). 75 Rib 9 (last sternal [true] rib – i.e. with direct sternal attachment). 76 Costal arch (fused costal cartilages of ribs 10–12 with fibrous connection to costal cartilage of rib 9). 77 Rib 13 (last or floating rib united with costal arch by fibrous tissue). 78 Intercostal space 6 (between bony ribs 6 and 7). 79 Interchondral space 6 (between costal cartilages of ribs 6 and 7). 80 Thoracic inlet (bounded by sternal manubrium, first pair of ribs and costal cartilages and thoracic vertebra 1). 81 Manubrium of sternum (sternebra 1 elongated into base of neck). 82 Sternebrae 2 and 3 (sternal segments joined by intersternebral cartilage 2). 83 Xiphoid cartilage of sternum (cartilaginous prolongation of last [8th] sternebra, xiphoid process, into belly wall).

Landmarks of appendicular skeleton

84 Dorsal (vertebral) border of scapula. 85 Cranial angle of scapula. 86 Caudal angle of scapula. 87 Scapular spine. 88 Acromion process of scapula. 89 Greater tuberosity of humerus (point of shoulder). 90 Iliac crest. 91 Sacral tuberosity of ilium (point of croup). 92 Coxal tuberosity of ilium (point of haunch). 93 Ischiatic tuberosity (point of buttock). 94 Greater trochanter of femur.
(D After Taylor, 1955)

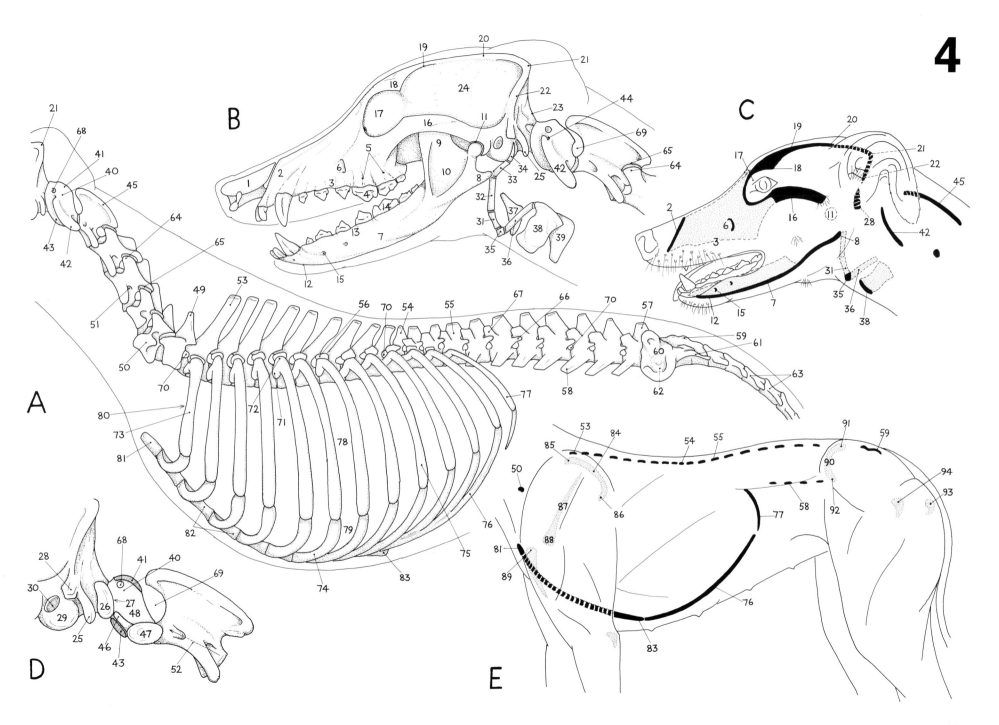

4

A

B

C

D

E

15

5

AXIAL SKELETON OF THE DOG – SKULL

The skull is given further coverage in this drawing because of its complexity. Drawn again from the left side in **A** the skull has now been 'pulled apart' somewhat: the jaw has been disarticulated and dropped down, the nasal and auricular cartilages have been removed from the skull and are figured separately, and the hyoid arch and larynx have been disarticulated from the cranium and pulled back slightly.

The drawing at the lower left (**B**) shows this disarticulated skull having been split longitudinally in the median plane, the drawing displaying the right half of the skull from the midline. This view does give considerable detail of the interior of the skull especially the extent of the cavities internally – nasal, paranasal, nasopharyngeal, and cranial, an aspect which we will deal with more fully in later drawings.

The head is large in relation to the body accommodating a large brain and well developed sense organs. However, the heads of different breeds vary enormously in conformation exhibiting a greater range of size and shape than in any other mammalian species. Head appearance is largely determined by skull shape, eye position and size, and form and carriage of the ears. On the right of the page the drawings shown in **C** and **D** illustrate the three skull categories that are generally recognized:

Mesaticephalic – a head of 'medium' proportions, neither unusually long nor too short, nor too wide or abnormally narrow, and found in a majority of breeds.

Dolichocephalic – long and narrow and found in such breeds as greyhounds, collies and borzois.

Brachycephalic – short and wide and found in such breeds as bulldogs, bostons and small spaniels.

Without going into detail about individual bones the skull can be simplified if we relate its structure to the functions that a dog uses its head for.

Firstly, the most important and potentially vulnerable part of the central nervous system, the brain, lies inside the head. The skull therefore provides housing and protection for it in the form of a 'brain-box' or **cranium**.

Secondly, in everyday life when moving around the head is normally the first part of the body which 'encounters the environment' consequently it contains the *major sense organs* responsible for collecting information from the surroundings. The skull provides capsules for housing and protecting these delicate organs:

Olfactory capsules – *ethmoid apparatus* supporting the nasal organs within the facial region and completing the cranial cavity in front.

Auditory capsules – *petrosal bones* containing the auditory organs embedded in the cranial wall ventrolaterally.

Eyes – visual organs are not enclosed in bony capsules since mobility is a prime requisite, but housed within *orbits* and protected by a *bony orbital margin*. They lie at the junction between cranial and facial regions.

Thirdly, the predominant method that a dog has of 'handling its environment' lies in the use of its head on the end of a mobile neck. The head therefore contains elements such as *teeth, jaws, lips and tongue* which can perform these 'handling' activities necessary in feeding, self-grooming and also in fighting. An **upper jaw** is fused to the undersurface of the face: a **lower jaw** is movably hinged with the braincase.

Fourthly, the head provides the initial parts of both digestive and respiratory pathways. Skull elements such as the **hyoid apparatus** support the tongue, pharynx and larynx, while other parts of the skull provide extensive areas for the attachment of jaw muscles.

These four functional units combine to produce a skull which is broadly considered as having **cranial** and **facial** parts. The cranial part is a relatively constant component based on the braincase and auditory organ capsules, with such concerns as jaw muscle attachment determining its overall architecture. The brain inside the cranium is directly continuous with the spinal cord contained within the vertebral canal along the length of the back. But, the cranium is a rigid box whereas the vertebral canal is contained within a series of jointed vertebral segments and its walls are mobile. The whole skull in fact is formed by an interlocking of numerous bones, the fibrous joints between them being termed **sutures**. When a dog reaches its full adult size these sutures are infiltrated with bone and individual bones fuse together to form a single structural unit. What this means is that during growth some change in shape of the skull is possible, a capacity which is lost as maturity is reached.

The spinal cord enters the braincase through the *foramen magnum*, a large hole in the rear face of the skull. A pair of occipital condyles flank the foramen and articulate with the first cervical bone (atlas) as the atlantooccipital joint already mentioned (fig 4D).

The brain has an extensive blood supply and gives rise to numerous nerves passing to and from all parts of the head. Since these nerves and blood vessels must pass through the walls of the brainbox to reach the brain, the cranium, particularly ventrolaterally, is perforated by several holes (foramina). The cranial cavity is shown in the midline section of the skull (**B**), and the position of the cavity in relation to the surface of the head is shown in fig 26.

Forming an important component of the braincase wall, and situated low down at the ventrolateral braincase margins towards the rear end of the skull, are the auditory capsules containing the membranous labyrinth of the inner ear (semicircular canals and cochlea). Each capsule is a compact mass, the *petrous temporal bone*, buried deeply inside the skull and best seen on inspection of a cut skull (**B**). A part is exposed on the caudal surface of the skull as the **mastoid process** between the external acoustic meatus and the occipital bone, behind the ear and palpable from the surface.

The sense of balance and equilibrium are perfectly well served by the sensory organs of the membranous labyrinth of the inner ear. However, the hearing (auditory) function requires an additional apparatus for transmitting sound waves from the head surface to the canals of the inner ear. Three minute bones, **ear ossicles** (malleus, incus and stapes) are situated at the ventrolateral aspect of the cranium and extend from the **eardrum** to an opening in the petrous temporal bone. These tiny bones are set in vibratory motion by sound waves impinging upon the eardrum, and convey this sensation to the fluid within the canals of the inner ear. To be able to vibrate, the ossicles must be isolated in an air-filled space (their very delicacy of form also means that they require protecting). To fulfil these two requirements they are enclosed in an ensheathing capsule of bone, the *tympanic bulla*, which fuses onto the surface of the braincase and encloses the air-filled **middle ear cavity**. Entering the cavity laterally immediately behind the jaw joint is the *external auditory canal* (acoustic meatus) across which the eardrum (**tympanic membrane**) would be stretched in life. The cavity retains a connection with the exterior through a membranous *auditory (pharyngotympanic or Eustachian) tube*. From an opening in the nasopharynx above the soft palate at the back of the throat (fig 26D) this short narrow tube leads into the middle ear cavity by passing dorsally and laterally into the tympanic bulla (fig 26C) and enables air pressure to be equalized on both sides of the eardrum. Pressure equalization occurs during swallowing since soft palate muscles are attached to and pull on its ventral wall opening the cavity of the tube which is normally closed. The tube is surrounded and supported by an inverted V-shaped cartilage which will allow little swelling should the tube wall be inflamed. Temporary blockage is thus always a possibility during throat infections.

The external aspect of the cranial region is especially influenced by the **pinna** of the external ear. This is the superficial apparatus for collecting and conducting sound inwards to the eardrum. A pinna lies to the side of the external occipital protuberance with the *opening of the ear canal* facing dorsolaterally. Pinna shape is variable − in ancestral dogs a highly mobile erect pinna was present, and erect ears are present in a number of modern breeds. In other breeds, whilst much of the pinna is erect its tip is pendulous, hanging down over the ear canal opening. At the extreme practically the entire pinna may be a large pendulous flap.

Like the nose the external ear is supported on a cartilaginous framework, the **auricular (conchal) cartilage**, a funnel-shaped and sufficiently stiff cartilage surrounding and maintaining the patency of an *ear canal* leading down to the tympanic membrane stretched across the external acoustic meatus. Two clear parts are recognized with the bend in lop-eared dogs where the two meet. An outer, wide open triangular leaf-like portion, the *scapha (helix)*, and a basal tubular rolled part surrounding the ear canal, the *concha*. The ear canal may be up to 7 cm long and begins at the entry into the concha, extends vertically downwards and then bends rostromedially through about 90° as a horizontal component leading to the eardrum. The rolled concha is palpable through the skin and intervening parotid salivary gland and its sharp right angled bend can be clearly appreciated. The canal can be straightened by careful manipulation, pulling the ear caudally, laterally and ventrally.

If you were to lift the helix of the pinna you could follow its caudal border down towards the opening of the ear canal. *En route* you will notice a *marginal cutaneous pouch*, a constant feature but one whose significance is unclear. At the caudal boundary of the opening into the ear canal you will feel this caudal border becoming thickened as the *antitragus*. A distinct *intertragic notch* separates the antitragus from the *tragus*, the thickened part of the concha forming the lateral boundary of the ear canal opening. Now return to the helix and follow its rostral border and you will feel that as it approaches the ear canal it subdivides. The large internal subdivision you will feel forming a prominent ridge on the internal wall of the opening into the ear canal overlapping the tragus medially. The lateral division is palpable at the rostromedial edge of the concha separated from the tragus by a *pretragic notch* in the rostral margin of the ear canal opening. Intertragic and pretragic notches are the 'landmarks' used by a veterinary surgeon should he want to perform a tragic resection for the relief of chronic otitis. The last palpable and visible feature in this area is the *anthelix*, a transverse ridge on the medial wall of the base of the helix near the entrance into the ear canal opposite the tragus. It is actually the free edge of a deep indentation on the outer convex surface and marks the division between helix and concha where the bend in lop-eared dogs occurs.

The skin covering the cartilage is more sparsely haired on the inner face of the funnel, and in the depths of the ear canal sebaceous and tubular ceruminous skin glands are numerous. The secretion from these glands combine to form ear wax (cerumen) which protects the delicate eardrum from harm. The pinna is provided with an extensive array of muscles at its base derived from facial muscles. All of the varied movements that the external ear can perform are brought about by these auricular muscles. An additional flattened L-shaped cartilage, the *scutiform*, lies on the surface of the temporal muscle rostromedial to the auricle. It functions to converge and redirect the pull from rostral auricular muscles.

The **facial part of the head** is a more variable component based on the jaws and upper parts of the respiratory and digestive tracts and attached to the rostral end and underside of the cranium. It contains the openings of the mouth, nose and those for eyes although its shape is most heavily influenced by jaws and teeth. The necessary elongation to provide an alveolar margin for tooth implantation produces a pointed rostral end to the face. Housing of olfactory organs on the other hand gives rise to a fairly wide and deep base blending with the front of the cranium. It is this facial part which in brachycephalic breeds is shortened and widened, whilst in dolichocephalic types it is lengthened and narrowed. In a number of breeds as well there is a discrepancy in length between upper and lower jaws. A short face is generally prognathic with an undershot lower jaw: a long face is often accompanied by a brachygnathic, receding lower jaw. Dogs with shortened faces as I'm sure you know may also have problems breathing through their noses and/or difficulties with the bite of their jaws.

The dorsal profiles of the cranium and face are in approximately parallel planes, the nasofrontal suture marking the boundary on the dorsal surface between the two. The step down from the cranium to the facial level is the nasofrontal angle or **stop**. In brachycephalic breeds the shortened, broadened face is coupled with a deepened stop and eyes that are directed more forwards. Cranial and facial regions in all breeds are joined on either side by a bar of bone, the **zygomatic arch**, which forms the outer boundary of an orbitotemporal fossa containing the eyeball and jaw muscle. The orbit therefore lies at the craniofacial boundary although normally considered a part of the face.

Bony basis of face and jaws
1–4 Nasal cartilages. **1** Internasal septum (cartilaginous septum). **2** Dorsolateral nasal cartilage. **3** Ventrolateral nasal cartilage. **4** Accessory nasal cartilage. **5** Nasal process of incisive bone (bordering nasal opening leading into bony part of nasal cavity). **6** Nasal bone. **7** Maxillary bone. **8** Dorsal nasal concha (nasoturbinate bone – supporting nasal mucous membrane). **9** Ventral nasal concha (maxilloturbinate bone – supporting nasal mucous membrane). **10** Nasal meatuses (air passageways through nasal cavity). **11** Nasopharyngeal meatus (airway leading back to internal nostrils). **12** Internal nostril (choana – leading into nasopharynx). **13** Cribriform plate of ethmoid bone (separating cranial from nasal cavities). **14** Ethmoidal labyrinth (ethmoturbinate bones – attached to cribriform plate and supporting nasal mucous membrane). **15** Maxillary recess (lateral diverticulum from nasal cavity). **16–17** Frontal sinus. **16** Lateral part of frontal sinus (large empty space in frontal bone). **17** Medial part of frontal sinus (containing extensions from ethmoidal labyrinth). **18** Hard palate (from palatine processes of incisive, maxillary and palatine bones). **19–28** Mandible (lower jaw). **19** Body of mandible. **20** Angular process of mandible. **21** Ramus of mandible. **22** Coronoid process of mandible (insertion of temporal muscle). **23** Masseteric fossa of mandible (insertion of masseter muscle). **24** Condyloid (articular) process of mandible (transverse to articulate with mandibular fossa of temporal bone in jaw joint). **25** Mandibular notch (separating coronoid and condyloid processes of mandible). **26** Mandibular fossa (smooth articular area for mandibular condyle in formation of jaw joint). **27** Retroarticular process (caudal boundary of mandibular fossa). **28** Mandibular symphysis (fibrocartilaginous intermandibular joint). **29** Alveolar borders of maxillary bone and mandible (tooth-bearing areas of upper and lower jaws).

Cranium and ears
30 Orbit (housing and protecting eyeball). **31** Zygomatic arch (bridge of bone connecting face and cranium below eye). **32** Zygomatic (supraorbital) process of frontal bone. **33** Temporal line (rostral divergence of external sagittal crest). **34** External sagittal crest (in dorsal midline of cranium). **35** External occipital protuberance (most dorsocaudal portion of cranium). **36** Nuchal crest (division between dorsal and caudal surface of cranium). **37** Parietal bone. **38** Frontal bone. **39** Temporal fossa (origin of temporal muscle). **40** Occipital bone. **41** Occipital condyle (both condyles form atlantooccipital joint with atlas). **42** Jugular process of occipital bone. **43** Foramen magnum (passage of spinal cord). **44** Cranial cavity (containing brain). **45** Bony cerebellar tentorium (separating cerebral from cerebellar hemispheres). **46** Hamulus of pterygoid bone. **47** Basisphenoid bone. **48–52** Temporal bone. **48** Petrous component of temporal bone (pyramid – housing membranous labyrinth of inner ear). **49** Tympanic bulla (surrounding tympanic [middle ear] cavity containing three auditory ossicles – incus, malleus and stapes). **50** External acoustic meatus (across which eardrum stretched in life). **51** Squamous part of temporal bone. **52** Mastoid process of temporal bone (sole representation on skull surface of petrous temporal bone). **53–67** Auricular cartilage. **53** Scapha (helix). **54** Apex. **55** Medial border of helix. **56** Spine of helix. **57** Lateral crus of helix. **58** Medial crus of helix. **59** Lateral border of helix. **60** Antitragus. **61** Lateral process of antitragus. **62** Exterior opening of vertical part of ear canal. **63** Concha. **64** Tragus. **65** Pretragic notch. **66** Intertragic notch. **67** Anthelix.

Skull foramina for nerve and blood vessel passage
68 Orbital fissure (rostral lacerate foramen – passage of oculomotor, trochlear and abducent nerves and ophthalmic branch of trigeminal nerve). **69** Foramen lacerum (middle lacerate foramen/internal carotid foramen – passage of internal carotid artery). **70** Jugular foramen (caudal lacerate foramen – passage of glossopharyngeal, vagus and accessory nerves and internal jugular vein). **71** Optic canal (passage of optic nerve). **72** Round foramen (passage of maxillary branch of trigeminal nerve – position indicated by broken circle in fig A in medial wall of alar canal). **73** Oval foramen (passage of mandibular branch of trigeminal nerve). **74** Hypoglossal foramen (passage of hypoglossal nerve). **75** Stylomastoid foramen (passage of facial nerve). **76** Condyloid canal passage of condyloid vein). **77** Lacrimal foramen (passage of nasolacrimal duct). **78** Infraorbital foramen (passage of infraorbital branches of maxillary nerve and vessels). **79** Alar canal of sphenoid wing (passage of maxillary artery). **80** Rostral alar foramen (exit from alar canal). **81** Caudal alar foramen (entry into alar canal). **82** Mandibular foramen (passage of mandibular alveolar nerve and vessels). **83** Mental foramina (passage of mental branches of mandibular alveolar nerve and vessels). **84** Canal for transverse venous sinus in occipital bone. **85** Canal for trigeminal nerve through petrous temporal bone. **86** Internal acoustic meatus (passage for vestibulocochlear [auditory] nerve). **87** Cerebellar fossa of petrous temporal bone (housing paraflocculus of cerebellum).

Hyoid apparatus supporting tongue and larynx, and laryngeal cartilages
88 Basihyoid bone (transverse element in floor of throat). **89–92** Cranial horn of hyoid (lying in pharyngeal wall). **89** Ceratohyoid bone. **90** Epihyoid bone. **91** Stylohyoid bone. **92** Tympanohyoid cartilage (articulating with tympanic bulla and mastoid process). **93** Caudal horn of hyoid (thyrohyoid bone in wall of laryngopharynx articulating with thyroid cartilage of larynx). **94** Epiglottic cartilage. **95** Thyroid cartilage. **96** Cricoid cartilage. **97** Arytenoid cartilage.

Teeth
I1–I3 Incisor teeth (3 upper and 3 lower). **C** Canine teeth. **P1–P4** Premolar teeth (4 upper and 4 lower teeth – with 'milk' precursors except for premolar 1). **M1–M3** Molar teeth (2 upper and 3 lower teeth – without 'milk' precursors).

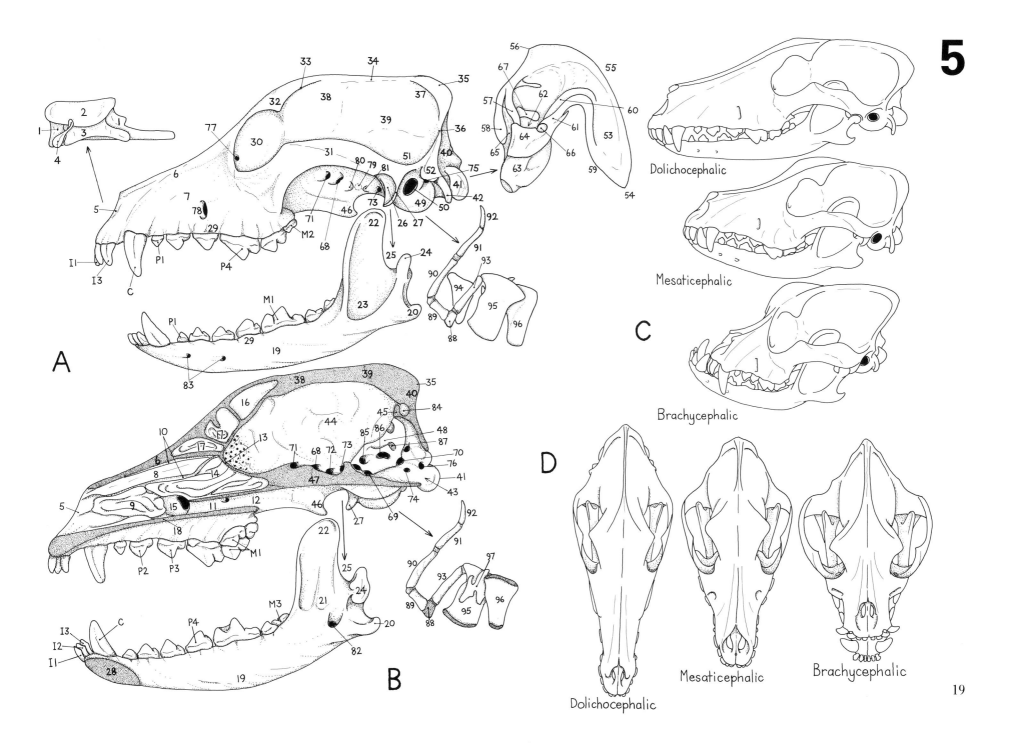

5

A

B

C

Dolichocephalic

Mesaticephalic

Brachycephalic

D

Dolichocephalic

Mesaticephalic

Brachycephalic

19

6

APPENDICULAR SKELETON OF THE DOG – FORELIMBS AND HINDLIMBS

The appendicular skeleton is shown in the drawings in the position that it might occupy in a standing dog. The main features of both forelimb and hindlimb skeletons are labelled on the drawings and those major bony points palpable from the surface are indicated on the accompanying surface views of the limbs. I do not mean to imply that these are the *only* bony prominences that you might feel; they represent significant 'landmarks' indicative of the position of the skeleton in the limbs. I'm sure that you could palpate much more than these on your own animal. For example, in the forelimb careful palpation through the overlying shoulder muscles will identify the cranial and caudal borders of the scapula, while practically the entire length of the humeral shaft can be felt through the upper arm muscles. Lower down in the limb close to the carpus much of the front surface of the radius is subcutaneous, and in fact if you were to lift up and manipulate the dog's paw you would be able to feel in the forearm both radius and ulna and how they move in relation to each other when you pronate (pads down) and supinate (pads up) the paw. Compare this with the greater mobility that you have in your own forearm. In the paw, carpal, metacarpal and phalangeal bones are all palpable through overlying tissues.

In the hindlimb little of the pelvic bone is palpable apart from the three tuberosities (sacral, coxal and ischiatic) because of the overlying rump and thigh musculature. These muscles also mean that the femoral shaft may be more difficult to feel than the humerus. Lower down the entire inner face of the tibia is subcutaneous and the bones of the hindpaw are all clearly palpable as in the forelimb.

It is quite reasonable to consider forelimbs and hindlimbs together since they are comparable in many respects. They both consist of four major segments:

1 **Pectoral (shoulder) girdle** containing the scapula (shoulder blade)/ **Pelvic (hip) girdle** containing the pelvic (hip) bone.

2 **Arm** (brachium) containing the humerus/**Thigh** containing the femur.

3 **Forearm (antebrachium)** containing the radius medially and ulna laterally/**Leg (crus** or **shank)** containing the tibia medially and fibula laterally.

4 **Forepaw** containing the bones of the 'hand'/**Hindpaw** containing the bones of the 'foot'.

Comparability is also evident within the paws themselves each consisting of three segments:

1 **Carpus** ('wrist') and **Tarsus** ('ankle' or hock) – both containing seven bones.

2 **Metacarpus** ('palm') and **Metatarsus** ('sole') – containing basically five metacarpal or metatarsal bones, although the first is almost always small or even absent altogether.

3 **Digits** ('toes') – typically five, each with three phalangeal bones. The first

digit ('dewclaw') in the forepaw is always small and only has two phalanges: the first digit in the hindpaw is usually absent although its presence is characteristic of certain breeds.

Although comparable in position and function the girdles joining limb with body differ quite significantly in structure. The **shoulder (pectoral) girdle** consists of a pair of **scapulae** (shoulder blades) lacking any bony connection with the skeleton of the trunk. This contrasts with the arrangement in ourselves in which well developed **clavicles** (*collar bones*) connect our shoulder blades with our sternum (breastbone). A very small collar bone is occasionally present in a dog cranioventral to the shoulder joint on a line between the sternal manubrium and the acromion process of the scapula although not attached to either. It would lie embedded in the *brachiocephalic muscle* as a narrow slip of predominantly fibrous tissue up to 1 cm long. Since it is poorly if at all ossified it does not normally show up even on X-ray pictures.

Lacking a collar bone allows the dog's scapula to have more mobility. It is held practically upright in the same vertical plane as the limb bones and can be thought of as the uppermost component of the limb skeleton moving as part of the limb. This flexible arrangement is due in some measure to the frontlegs being used as shock absorbers as well as propulsive units, the landing shock being taken up by those muscles and tendons holding the scapula in place.

The **hip (pelvic) girdle** consists of a pair of **hip bones** each one forming

from three separate bones (*ilium, ischium and pubis*) which fuse early in life. (Strictly speaking there are four bones making up a hip bone, the fourth is a small *acetabular bone* contributing to the acetabular cup of the hip joint.) Each half of the girdle therefore appears as a single hip bone united with its fellow of the opposite side in the midventral *pelvic symphysis*. The two halves are also firmly joined with the sacrum on either side through robust *sacroiliac joints* which allow little appreciable movement. The joints are strengthened above and below by bands of strong fibrous tissue, the sacroiliac ligaments (see fig 7F). Thus a fairly rigid box-like framework open at the sides is produced, able to withstand considerable force as it transmits thrust from the hindlimbs to the backbone during locomotion.

The corresponding joint in the shoulder girdle to the sacroiliac is a highly mobile 'muscular joint' based on a fan-shaped *ventral serrate muscle* attached between the upper end of the scapula and the ribs and cervical vertebrae. The muscles of either side form a sling suspending the trunk between the limbs. Some of the possible reasons for the presence of a muscular rather than a bony union will be explained later.

The **humerus** (arm bone) and **femur** (thigh bone) are substantial long bones of comparable shape and function. The humerus extends from a head articulating with the glenoid cavity of the scapula at the shoulder joint, to a condyle articulating with both radius and ulna at the elbow joint. The femur extends from a head situated in the acetabulum of the hip bone at the hip joint, to condyles articulating with the tibia at the stifle joint. Both bones have distinctly palpable proximal enlargements for muscle attachment — tubercles of the humerus and trochanters of the femur. In addition the femur has several sesamoid bones associated with it of particular note being the ovoid **patella** (knee cap) which forms in the tendon of the quadriceps femoris muscle at the front of the thigh and slides up and down in a groove on the lower end of the femur. A pair of stifle sesamoid bones (**fabellae**) are also located in the tendons of origin of the gastrocnemius muscle where these pass over the rear of the femoral condyles.

Shoulder and **hip joints** are both ball-and-socket joints although the glenoid cavity of the shoulder is shallow while the acetabular fossa of the hip is a deep cup-shaped cavity. The femoral head sits quite snugly inside this cup suggesting that the hip is a more stable joint than the shoulder. Nevertheless, because of its strong surrounding muscles and its more limited range of movements, the shoulder joint rarely dislocates.

Both forearm bones are well developed: the **radius**, of more or less uniform size throughout its length, is the main weight supporter transferring weight from humerus to paw; the **ulna** is a tapering bone well developed at the elbow where it forms most of the joint, as well as providing a lever (*olecranon process — point of the elbow*) behind the joint for the attachment of its extensor muscles. Both bones articulate with the humerus at the elbow and with the proximal row of carpal bones at the 'wrist' restricting movement between them as we have already mentioned. The **elbow joint** although quite complex in structure, functions primarily as a hinge. The humeral condyle articulates with both the trochlear notch of the ulna and the head of the radius. The *humeroulnar articulation* is the major 'stabilizing component' restricting elbow movement to a single plane — flexion and extension. The *humeroradial articulation* transfers weight from forearm to upper arm. A third articulation between the radial head and an ulnar notch allows for a slight degree of rotation of the radius in relation to the ulna. Since the radius forms the main articulation with the paw at the carpus, rotation at the radioulnar component of the elbow joint turns the paw inwards (pronation) and outwards (supination). However, rotation in the forearm, and hence rotation of the paw, is limited in the dog, in comparison with ourselves, because of the contributions of radius and ulna to both elbow and carpal joints.

The **tibia** in the leg is by far the larger of the two components being the sole articulation with the femur at the stifle joint, whilst lower down it forms the articulation with the trochlea of the talus at the hock joint. The slender **fibula** only articulates with the tibia and is retained since it still serves for muscle and ligament attachment. The **stifle joint** in the hindlimb is another exceedingly complex joint. Like the elbow it also consists of more than one joint component and also operates mainly as a 'hinge', flexing and extending. It is made up of a *femorotibial joint* between knuckle-shaped femoral condyles and flattened tibial condyles, and a *femoropatellar joint* between the patella and a groove on the front of the lower end of the femur.

The **carpus** (wrist) and **tarsus** (ankle or hock) are terms used to designate a topographical region, a composite of joints and a collection of bones. Each is composed of seven short bones although not all are strictly comparable between the two. An upper (proximal) row is represented in the carpus by a *radial carpal bone* on the medial side and an *ulnar carpal bone* laterally: in the tarsus by the *talus* (astragalus or tibial tarsal bone) on the medial side and the *calcaneus* (fibular tarsal bone) laterally. A lower (distal) row has four *carpal/tarsal bones* articulating with the metacarpals or metatarsals below. The seventh bone in the hock is a medially positioned *central tarsal bone* sandwiched between talus above and tarsals 1–3 below. The seventh wrist bone, the *accessory carpal*, is an extra element projecting back from the outer end of the upper carpal row to provide attachment for forearm muscles which prevent carpal overextension and collapse. A somewhat similar arrangement exists in the hock but here the backwardly projecting process (*calcaneal tuberosity — point of the hock*) is not a separate bone but merely an extension of the calcaneus to which hock extensor muscles attach.

The **carpal** and **tarsal joints** are both composite series of articulations between the constituent bones, as well as between these bones and the adjoining segments of the limbs. Despite this complexity both function essentially as hinges — flexing and extending, and this movement takes place primarily at the main joint surface: an **antebrachiocarpal joint** between the radius and ulna and the radial and ulnar carpal bones; a **talocrural** joint between the tibia and the talus and calcaneus.

The elongated **metacarpals** or **metatarsals** support the digital bones (**phalanges**). The skeleton of each fully developed digit (digits 2–5) consists of three phalanges, the distal ones being modified in shape according to the shape of the horny claw in which they are enclosed. Digits also contain several sesamoid bones associated with the joints between the metacarpals or metatarsals and the proximal ends of the first phalanges — **metacarpophalangeal/ metatarsophalangeal joints**. At the rear of each joint a pair of *proximal sesamoid bones* is situated in the tendon of insertion of the *interosseous muscle* of that digit. These bones actually form an important joint component held in position by a number of ligaments (see fig 7D). As well as combating the friction to which the tendons are subjected, they also form a pulley over which digital flexor tendons pass on their way down into the digits. When a dog is standing up or when it is taking weight on its paw when moving around, the metacarpophalangeal and metatarsophalangeal joints are in an 'overextended' condition. Normally flexing or folding of a joint involves movement of one bone in relation to another so as to reduce the angle between them. Conversely extension increases this angle up to 180°. However, metacarpophalangeal and metatarsophalangeal joints extend beyond 180° in a normal standing position, hence the term overextended. Beneath these joints lies the large metacarpal or metatarsal pad, they therefore accept the weight of the body pressing down onto the pad.

The **phalangeal joints**, *proximal and distal interphalangeal* in each digit, are closely comparable in forepaw and hind-paw and again are functional if not strictly anatomical hinges, flexing and extending. The angles at these digital joints in a normal standing position are quite marked: the proximal inter-phalangeal joint has a ventral angle of about 135°, the dorsal angle of the distal interphalangeal joint is about 90°. The digits are therefore not splayed out flat on the ground as many skeletal drawings and museum preparations would have us believe.

Topographical regions of forelimb
1 Scapular region (shoulder). **2** Shoulder joint region. **3** Brachial region (arm or upper arm). **4** Tricipital region (caudal border of arm based on tricipital muscle). **5** Cubital region (elbow). **6** Olecranon region. **7** Antebrachial region (forearm). **8** Carpal region (wrist). **9** Metacarpal region (fore pastern). **10** Phalan-geal region (digits or toes designated by roman numerals – I = dewclaw or 1st digit).

Forelimb skeleton
11–22 Scapula (shoulder blade – pectoral girdle). **11** Spine of scapula. **12** Supraspinous fossa of scapula. **13** Infraspinous fossa of scapula. **14** Acromion process of scapula. **15** Dorsal (vertebral) border of scapula. **16** Caudal angle of scapula. **17** Caudal border of scapula. **18** Infraglenoid tubercle. **19** Cranial angle of scapula. **20** Cranial border of scapula. **21** Scapular notch at level of neck of scapula (constricted region separating off articular angle in which lies glenoid cavity for humeral head). **22** Supraglenoid tubercle of scapula. **23–34** Humerus (arm bone). **23** Head of humerus (larger than glenoid cavity of

scapula). **24** Neck of humerus (only distinct caudally). **25** Greater tubercle of humerus (point of shoulder). **26** Crest of greater tubercle of humerus. **27** Tricipital line of humerus. **28** Deltoid tuberosity of humerus. **29** Brachial (musculospiral) groove of humeral shaft (begins on caudal surface of humeral neck and ends on cranial surface of shaft). **30–31** Condyle of humerus. **30** Capitulum of humeral condyle (articulating with head of radius). **31** Trochlea of humeral condyle (larger than capitulum and pulley-shaped, articulating with trochlear notch of ulna). **32** Supratrochlear foramen of humerus. **33** Medial (flexor) epicondyle of humerus. **34** Lateral (extensor) epicondyle of humerus. **35–44** Forearm bones. **35–39** Radius. **35** Head of radius (articulating with capitulum of humerus and radial notch of ulna). **36** Neck of radius. **37** Lateral tuberosity of radius. **38** Body (shaft) of radius. **39** Medial styloid process of radius. **40–43** Ulna. **40** Olecranon process of ulna (point of elbow). **41** Trochlear (semilunar) notch of ulna. **42** Body (shaft) of ulna. **43** Lateral styloid process of ulna (articu-lating with ulnar carpal and accessory carpal bones). **44** Antebrachial interosseous space between radius and ulna. **45–56** Forepaw bones. **45–48** Carpal bones. **45** Radial carpal bone (scaphoid). **46** Ulnar carpal bone (tri-quetrum). **47** Accessory carpal bone (pisi-form). **48** Carpal bones 1–4 (trapezium, trapezoid, capitate and hamate). **49–51** Meta-carpal bones. **49** Metacarpal bone 1. **50** Meta-carpal bone 2. **51** Lateral surface of base of metacarpal bone 5. **52–56** Digital bones. **52** Proximal (1st) phalanx of digit 2. **53** Middle (2nd) phalanx of digit 2. **54** Distal (3rd) phalanx of digit 2. **55** Ungual process of 3rd phalanx. **56** Proximal sesamoid bones of metacarpophalangeal joint of digit 5 (pair of sesamoids associated with each metacarpo-phalangeal joint).

Joints of forelimb
57 Shoulder joint. **58–59** Elbow joint. **58** Humeroulnar component of elbow joint. **59** Humeroradial component of elbow joint. **60** Antebrachiocarpal joint (main component of carpal joint). **61** Metacarpophalangeal joint of digit 2. **62** Proximal interphalangeal joint of digit 5. **63** Distal interphalangeal joint of digit 5.

Topographical regions of pelvis and hindlimb
64 Gluteal region (rump). **65** Coxal tuberosity region (haunch). **66** Clunial region (including ischiorectal fossa). **67** Ischiatic tuberosity region (buttock). **68** Hip joint region. **69** Femoral region (thigh or upper thigh). **70** Genual region (stifle). **71** Patellar region. **72** Popliteal region. **73** Crural region (leg, shank or lower thigh). **74** Tarsal region (ankle or hock). **75** Calcaneal region. **76** Metatarsal region (rear pastern).

Hindlimb skeleton
77–88 Hip bone (pelvic bone of pelvic girdle). **77–84** Ilium of hip bone. **77** Crest of ilium (cranial border of ilium). **78** Wing of ilium. **79** Coxal tuberosity of ilium (point of haunch – cranial ventral iliac spine). **80** Caudal ventral iliac spine. **81** Sacral tuberosity of ilium (point of croup – cranial dorsal iliac spine). **82** Caudal dorsal iliac spine. **83–86** Ischium of hip bone. **83** Ischiatic tuberosity (point of buttock). **84** Ischiatic spine. **85** Greater ischi-atic notch of hip bone. **86** Lesser ischiatic notch of hip bone (converted into lesser ischi-atic foramen by sacrotuberous ligament). **87** Pecten of pubic bone (attachment area for cranial pubic ligament and prepubic tendon). **88** Obturator foramen (blocked in by obtu-rator membrane in life and perforated by obturator nerve and vessels). **89–99** Femur (thigh bone). **89** Head of femur (nearly hemi-spherical and fitting into deep cup-shaped

acetabulum of hip bone). **90** Neck of femur. **91** Greater trochanter of femur (point of hip). **92** Body (shaft) of femur. **93** Lateral supra-condylar tuberosity of femur. **94** Lateral epi-condyle of femur supporting a sesamoid fabella. **95** Lateral condyle of femur. **96** Medial condyle of femur (smaller and less convex than lateral). **97** Fabella supported on medial epicondyle of femur. **98** Femoral trochlea with lateral and medial trochlear ridges. **99** Extensor fossa of femur (origin of long digital extensor muscle). **100** Patella ('knee cap' – sesamoid bone in tendon of insertion of quadriceps femoris muscle and occupying trochlea of femoral condyle [patellar groove]). **101–110** Crural bones (bones of leg or shank). **101–107** Tibia. **101** Medial condyle of tibia. **102** Lateral condyle of tibia. **103** Ex-tensor groove of tibia (for passage of long digital extensor tendon). **104** Tibial tuberosity (insertion of patellar tendon). **105** Cranial border of tibia (tibial crest). **106** Body (shaft) of tibia. **107** Medial malleolus of tibia. **108–110** Fibula. **108** Head of fibula. **109** Body (shaft) of fibula. **110** Lateral malleolus of fibula. **111–118** Hindpaw bones. **111–116** Tarsal bones. **111** Talus (astragalus or tibial tarsal bone). **112** Trochlea of talus (articulatory surface with tibia). **113** Calcaneus (os calcis or fibular tarsal bone). **114** Calcaneal tuber-osity (point of hock). **115** Central tarsal bone (navicular). **116** Tarsal bones 1–4 (medial, intermediate and lateral cuneiform and cu-boid). **117–118** Metatarsal bones. **117** Meta-tarsal bone 1. **118** Metatarsal bone 5 (lateral surface, base).

Joints of hindlimb
119 Hip joint. **120–121** Stifle joint. **120** Femoropatellar component of stifle joint. **121** Femorotibial component of stifle joint. **122** Crurotarsal joint (main component of tarsal joint). **123** Metatarsophalangeal joint, digit 2.

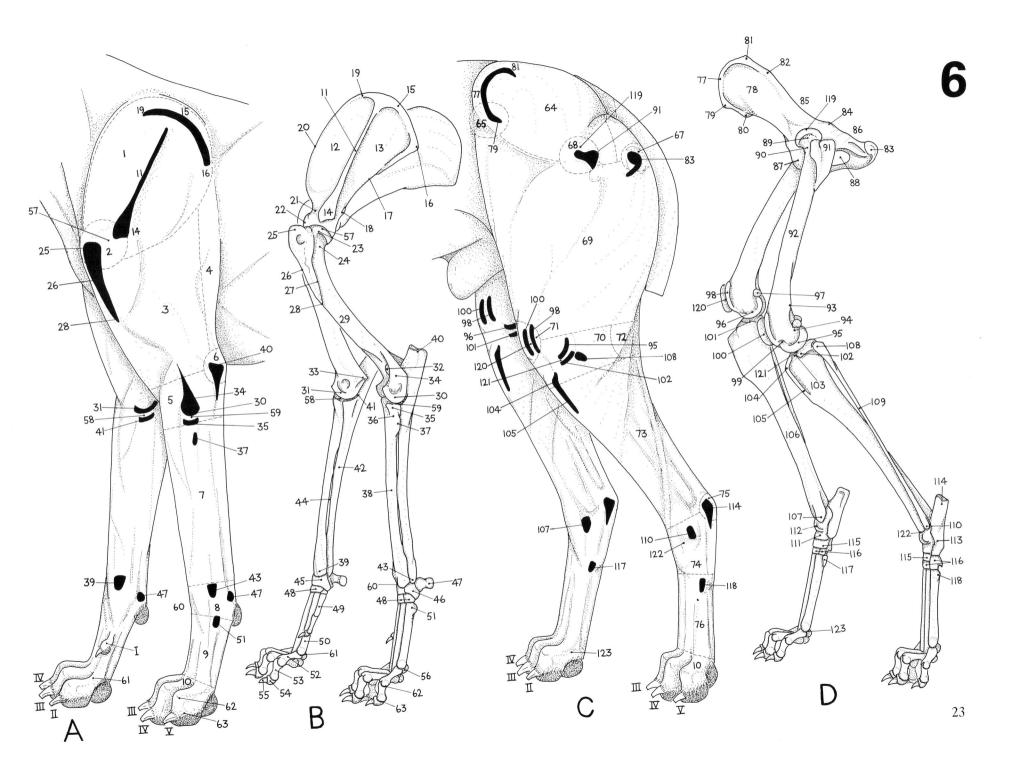

6

23

7

JOINTS AND LIGAMENTS OF THE SKELETON OF THE DOG

The main illustration, a lateral view of the skeleton, more or less duplicates that shown in fig 3 but with the addition of those ligaments (solid black areas) holding bones together. The inset sketches around the periphery of the main drawing are enlargements of several of the joints showing their associated ligaments in more detail.

This drawing clearly shows how a dog's skeleton is made up of numerous bones connected at joints to give an integrated structure. Thus it would seem that the basic function of a joint is, as its name suggests, to join bones together. Nevertheless, some also allow movement to occur between bones. How much movement is allowed will obviously depend upon the type and amount of tissue forming the basis of a joint. Should the union be formed of dense fibrous connective tissue or cartilage then the joint is likely to be an immovable one, or at best one which has only a limited range of movement.

Included in a completely immovable category of joints are the **cartilaginous growth plates** of developing bones. These simply join together growing centres of ossification and provide the cartilage for replacement by bone. Once a bone has attained its adult size growth plates are invaded by bone tissue so that separate centres of ossification fuse, obliterating joints.

In a slightly more mobile category of joints are **sutures**, the fibrous unions between skull bones which allow them to grow at their edges, altering in size and shape as a skull enlarges. Sutures also allow slight distorting movements to take place during birth assisting in the passage of a pup's head through the narrow confines of its mother's pelvis.

Both of the examples used so far (cartilaginous growth plates and fibrous sutures) are not permanent joints throughout life; they become invaded by bone and obliterated at some stage of growth. On the other hand **intervertebral discs**, the strong fibrocartilaginous joints firmly attached to and joining vertebral bodies, are permanent structures throughout life. Present in every intervertebral space except the first two in the neck, intervertebral discs provide probably the most important component keeping the backbone together. Dense fibrocartilage is a tissue which is strong enough to prevent vertebral bodies from either being pulled completely apart or being squashed too tightly together when the backbone is supporting weight and transmitting thrust, whilst at the same time still sufficiently flexible to allow some movement. The movements permitted by intervertebral discs are therefore squashing together and pulling apart of vertebral bodies, shortening and lengthening the spine, together with limited angular movement – flexion and extension of the spine in both vertical and horizontal planes. Joint movement is brought about by muscles contracting to overcome the resistance offered by the fibres in the cartilage of a disc. When these muscles relax and the moving force ceases the vertebrae return to a neutral position through the release of tension built up in the fibres of the disc.

An intervertebral disc actually has a 'soft centre', the *pulpy nucleus*. This serves to distribute forces evenly through the surrounding fibrous area of the disc when the joint is moved, and absorbs much compression when the joint is compressed. This soft centre has a tendency to harden as the dog ages and consequently cannot perform its functions to the same degree so that extra strain falls on the outer fibrous area and damage to it might occur. Disc damage may allow the somewhat hardened nucleus to infiltrate into the outer fibrous component weakening it further. Extrusion of the pulpy nucleus may reach the outer surface of the disc – so called 'slipped disc', and give rise to problems should it impinge upon nerves or the spinal cord. Disc damage is most likely to occur where there is greatest intervertebral movement, for instance in the lumbar region, and may well produce symptoms of pain, weakness or even paralysis because of its effects on nerves to the hindlimbs.

Joints displaying wider ranges of movement do not have a solid mass of tissue joining the component bones together, an intervening tissue layer is almost completely absent. Although the articular surfaces of bones are separated, the bones are still connected by a fibrous sleeve around a joint which merges with the periosteum of the bones at the periphery of their articular surfaces. This sleeve or capsule encloses a cavity containing a lubricant (synovial fluid) which reduces the friction between contacting surfaces as they move over one another. Such **synovial joints** are present throughout the body and include all limb joints.

Inside a synovial joint the contact surfaces of the bones are covered by a layer of *articular cartilage* which, unlike bone, has the necessary elasticity and compression capability to protect bone from fracturing by absorbing concussion. Cartilage is also useful in this situation because it lacks a nerve supply, hence the nervous system is not continually bombarded by information whenever a joint moves. Information concerning joint position and movement originates in sense receptors situated in the capsule of fibrous tissue around a joint. This capsule also supports a membrane on its internal surface responsible for secreting lubricant *synovial fluid* into the joint cavity. Although synovial fluid minimizes friction it does not prevent the bearing surfaces of articular cartilage from slowly being worn away. Thus just like other body fluids it contains scavenging cells which can destroy the minute bits and pieces of cartilage that are worn off of the articular surfaces during the habitual wear and tear of normal joint movement. Fortunately the cartilage is continually renewed by growth from within so that a bearing surface is always maintained. Synovial fluid also requires renewal, therefore it is continuously secreted into the joint cavity and continuously removed from it, circulating just like any other body fluid. Should the synovial membrane be inflamed, possibly as a result of infection or mechanical damage, upset of joint fluid circulation might result. Excess fluid could build up in the joint cavity, separating the working surfaces, stretching the joint capsule with resultant pain.

The extent and range of movements that a synovial joint may perform depend in large measure on the shape of the bones forming it. However, the fibrous layer of the sleeve-like capsule may also be locally thickened as **ligaments** (many of which are shown in the drawings in solid black shading), strong, practically inelastic structures, strategically positioned to prevent excessive or abnormal movements from occurring, whilst allowing normal or habitual movements to take place. Stretching and tearing of a ligament may occur if a joint is forced beyond its normal limits, the joint being *sprained*. Should ligament damage be more extensive it might allow the bones entering into a joint to be displaced relative to each other, the joint being *dislocated*. Both occurrences will be painful and, particularly the latter, may immobilize a joint.

Several categories of synovial joint can be recognized differing in the shape of their articular surfaces and the range of movement that they perform. **Ball-and-socket joints** like the *hip joint*, with its rounded articular head nestling in a cup-shaped cavity, will potentially allow a large range of movement: flexion and extension, abduction and adduction, and rotation. But, some possible movements at a joint may be limited by ligaments and muscles. Thus the ball-and-socket *shoulder joint* apparently has a wide range of movement, although in normal circumstances its movement is confined by surrounding muscles to limited degrees of flexion and extension. The

more extensive movements of abduction and adduction which appear to be made at the shoulder are actually movements of the whole limb and girdle on its muscular attachments to the trunk.

All of the joints lower down the limbs are restricted in their range of movements. The *humeroulnar component of the elbow joint* (**A**) is a true anatomical **hinge** with a spool-like humeral condyle fitting snuggly into a concave trochlear facet of the ulna. Here the shape of the bones has been modified to restrict movements to one plane only. Irrespective of their anatomical shape nevertheless, most of the joints in the lower parts of the limbs function as simple hinges, flexing and extending. A significant contributory factor in restricting joints to a hinge action is the presence of **collateral ligaments**, thickenings of the fibrous layer of a joint capsule on its medial and lateral aspects. As you can no doubt see in the accompanying drawings such collateral ligaments are characteristic features of all of the lower limb joints, supporting the suggestion that they all function as hinges.

Restricting movement to one of rotation of a bone around its own long axis is characteristic of **pivot joints** like the joint between the radial head and the ulnar notch at the elbow. Similar in function, although slightly different in structure, is the *atlantoaxial joint* between the atlas and axis vertebrae (see fig 4 D). Here the odontoid process of the axis projects forwards into the ring-shaped base of the atlas where it is confined in position by ligaments. The atlas rotates around the articulation with this axis peg — shaking the head.

Many joints have more or less flat

articulatory surfaces, consequently they do not show angular movements but display sliding movements in the same plane as the articulatory surfaces. **Plane joints** like this are found between the bones at the wrist and ankle regions where fairly restricted amounts of movement are possible. Despite the presence of numerous plane joints the overall movement at both carpus and tarsus is still predominantly hinge-like. In fact the main axis of movement in these areas is at the *antebrachiocarpal joint* between the forearm and wrist, and the *crurotarsal joint* between the lower leg and tarsus. The major ligaments of both joints are collaterals on medial and lateral sides.

Some synovial joints contain extra pieces of cartilage inside them, for instance the fibrocartilages (**meniscal cartilages**) of the *femorotibial component of the stifle (knee) joint* (**E** & **G**). These paired, C-shaped cartilages are ingrowths into the joint cavity from its capsule. Their undersurfaces are flattened and rest on the flattened tibial condyles, while their upper surfaces are concave to conform to the rounded femoral condyles. In some complex fashion they improve the efficiency of joint lubrication and may also serve to diminish concussion.

The stifle joint functions as a hinge with prominent collateral ligaments placed medially and laterally. It appears, however, to show gliding movements like a plane joint, the rounded femoral condyles moving backwards and forwards on the flattened tibial condyles as the joint flexes and extends. In order to limit these gliding movements and improve joint stability there is an ad-

ditional pair of ligaments right inside the joint joining femur to tibia. They cross over one another, viewed from the side, hence the term **cruciate ligaments** used for them (**G**), and during movement prevent the tibia from being displaced too far forwards or backwards in relation to the femur.

The stifle joint has now been mentioned several times because it has such a complex structure. Associated with it is a large sesamoid bone, the patella (knee cap), making up the *femoropatellar joint*. The patella runs up and down in a groove (femoral trochlea) at the lower end of the femur, and is formed within the tendon of the large stifle extensor muscle at the front of the thigh, the quadriceps femoris muscle. During stifle flexion and extension the patella slides up and down within its groove, maintained in its position by collateral ligaments extending to the femur both medially and laterally (**E**).

Although ligaments play an important role in joint stability a further very important stabilizer for any joint is the musculature that surrounds it. In fact muscles are in the advantageous position of being able to adjust in length and tension throughout the movement that they produce or that they resist. This is impossible for inelastic ligaments. It would seem quite possible that muscles are as important as ligaments in joint stabilization. The cooperative action of muscles in producing joint movement and stabilization will be considered later on.

The drawing also includes ligaments found in other parts of the body which are not necessarily associated with synovial joints but still help to connect bones and to prevent dislocation. The backbone, for instance, has a variety of ligaments, a particularly important one being the **supraspinous ligament** joining the tips of spinous processes together and reducing the possibility of excessive separation of vertebral bodies. In the neck it is modified as the **nuchal ligament** containing an extra component of elastic tissue. Passing forwards from the summit of the first thoracic spine to the spine of the axis this assists in head support and, due to its elastic nature, helps to raise the head and neck on the thorax reducing the muscular effort needed.

A further important ligament, the **sacrotuberous ligament**, lies in the pelvic wall (E). It is a stout band of fibrous tissue running from the ischiatic tuberosity (point of the buttock) to the sacrum and caudal vertebra 1, and provides attachment for some of the muscles of the rump and thigh. Below it blood vessels and nerves to the pelvic wall and hindlimb pass in and out of the pelvis over the surface of the pelvic bone through the greater and lesser ischiatic foramina. It will also therefore prevent compression of these vessels and nerves by surrounding muscle contractions.

Joints and ligaments of skull
1 Orbital ligament (completing orbital rim caudally joining zygomatic process of frontal bone with frontal process of zygomatic bone of zygomatic arch). **2** Jaw (temporomandibular) joint (synovial, hinge, containing an intra-articular cartilage subdividing joint cavity into upper and lower compartments). **3** Lateral ligament of temporomandibular joint. **4** Temporohyoid joint (fibrous, limited degree of movement). **5** Mandibular symphysis (fibrocartilaginous intermandibular joint, little if any movement).

Joints and ligaments of vertebral column
6 Intervertebral discs (fibrocartilaginous, very restricted movement at individual discs although sum total throughout column is quite considerable). **7** Zygapophyseal joints between cranial and caudal articular processes (synovial, some angular movement, restrict rotation between vertebrae). **8** Intertransverse ligaments (only distinct in lumbar region). **9** Interspinal ligaments. **10** Supraspinous ligament (uniting summits of spinous processes in trunk). **11** Nuchal ligament (continuation of supraspinous ligament in neck attaching cranially to spine of axis). **12** Lumbosacral joint (intervertebral joint between lumbar vertebra 7 and sacrum). **13** Atlantooccipital joint (synovial, hinge in action, flexion and extension of head on neck – 'yes' joint). **14** Lateral atlantooccipital ligament. **15** Atlanto-axial joint (synovial, rotation of head on neck – 'no' joint). **16** Dorsal atlantoaxial ligament or membrane.

Joints and ligaments of ribcage
17–18 Costovertebral joint. **17** Rib head (capitular) articulation with vertebral body (synovial, hinge in action). **18** Rib tubercle articulation with transverse process (synovial, hinge in action). **19** Ligament of rib head. **20** Ligament of rib tubercle (costotransverse ligament). **21** Sternocostal joint (synovial, hinge in action). **22** Dorsal sternal ligaments. **23** Ventral sternal ligaments (merging with sternal membrane). **24** Costochondral joints (fibrous, little appreciable movement).

Joints and ligaments of forelimb
25 Shoulder joint (synovial, ball-and-socket although movement restricted primarily to a hinge action by muscles such as supraspinatus and subscapular acting as collateral ligaments). **26** Articular capsule (capsular ligament) of shoulder joint. **27–29** Elbow joint (composite synovial, hinge in overall action). **27** Humeroulnar joint. **28** Humeroradial joint. **29** Proximal radioulnar joint. **30–31** Collateral ligaments of elbow joint. **30** Medial (radial) collateral elbow ligament. **31** Lateral (ulnar) collateral elbow ligament. **32** Annular ligament of radial head. **33** Oblique ligament of elbow joint. **34** Antebrachial interosseous ligament. **35** Distal radioulnar joint (synovial, contributing to forearm rotation). **36** Radio-ulnar ligament. **37–39** Carpal joints (composite, synovial, several plane joints although a hinge in overall action). **37** Antebrachio-carpal joint (carpal component at which most movement occurs). **38** Intercarpal joints. **39** Carpometacarpal joints. **40–41** Collateral ligaments of carpus. **40** Medial (radial) collateral carpal ligaments. **41** Lateral (ulnar) collateral carpal ligaments. **42** Dorsal intercarpal ligaments. **43** Palmar intercarpal ligaments. **44** Ligaments of accessory carpal bone. **45** Palmar carpal fibrocartilage. **46** Metacarpophalangeal joints of forepaw/metatarsophalangeal joints of hindpaw (synovial, hinge in action). **47** Medial and lateral collateral ligaments of metacarpophalangeal/metatarsophalangeal joints. **48** Medial and lateral collateral ligaments of proximal sesamoid bones. **49** Proximal interphalangeal joint (synovial, hinge in action). **50** Collateral ligaments of proximal interphalangeal joints. **51** Distal interphalangeal joint (synovial, hinge in action). **52** Collateral ligaments of distal interphalangeal joint. **53** Dorsal elastic ligaments.

Joints and ligaments of hindlimb
54 Sacrotuberous ligament (joining lateral sacral crest and transverse process of caudal vertebra 1 with ischiatic tuberosity). **55** Sacro-iliac joint (combined synovial and fibrocartilaginous, little or no movement). **56** Dorsal sacroiliac ligament. **57** Pelvic symphysis (combination of pubic and ischiatic symphyses – fibrocartilaginous, little if any movement). **58** Hip joint (synovial, ball-and-socket with wide potential range of movement). **59** Articular capsule (capsular ligament) of hip joint. **60–62** Stifle (knee) joint (complex synovial, hinge in action). **60** Femorotibial joint. **61** Femoropatellar joint. **62** Proximal tibiofibular joint. **63–64** Intraarticular fibrocartilages (semilunar meniscal cartilages) of stifle joint. **63** Lateral meniscus. **64** Medial meniscus. **65** Meniscofemoral ligament (femoral ligament of lateral meniscus). **66–67** Intraarticular ligaments (cruciate ligaments) of stifle joint (joining tibia and femur in interior of stifle joint). **66** Cranial (lateral) cruciate ligament. **67** Caudal (medial) cruciate ligament. **68–69** Collateral ligaments of stifle joint. **68** Medial (tibial) collateral stifle ligament. **69** Lateral (fibular) collateral stifle ligament. **70** Patellar ligament (continuation of tendon of quadriceps femoris muscle onto tibial tuberosity). **71–72** Femoropatellar ligaments (collateral ligaments of femoropatellar joint). **71** Medial femoropatellar ligament. **72** Lateral femoropatellar ligament. **73** Ligaments of fibular head. **74** Distal tibiofibular joint (synovial, immobile). **75** Tibiofibular ligaments. **76–78** Tarsal joint (composite synovial, hinge). **76** Talocrural joint (synovial saddle-shaped joint at which most tarsal movement occurs). **77** Intertarsal joints. **78** Tarsometatarsal joints. **79–80** Collateral ligaments of tarsal joint. **79** Medial collateral tarsal ligament. **80** Lateral collateral tarsal ligament. **81** Dorsal tarsal ligaments. **82** Long plantar ligament.
(E, F & G After Evans & Christensen, 1979)

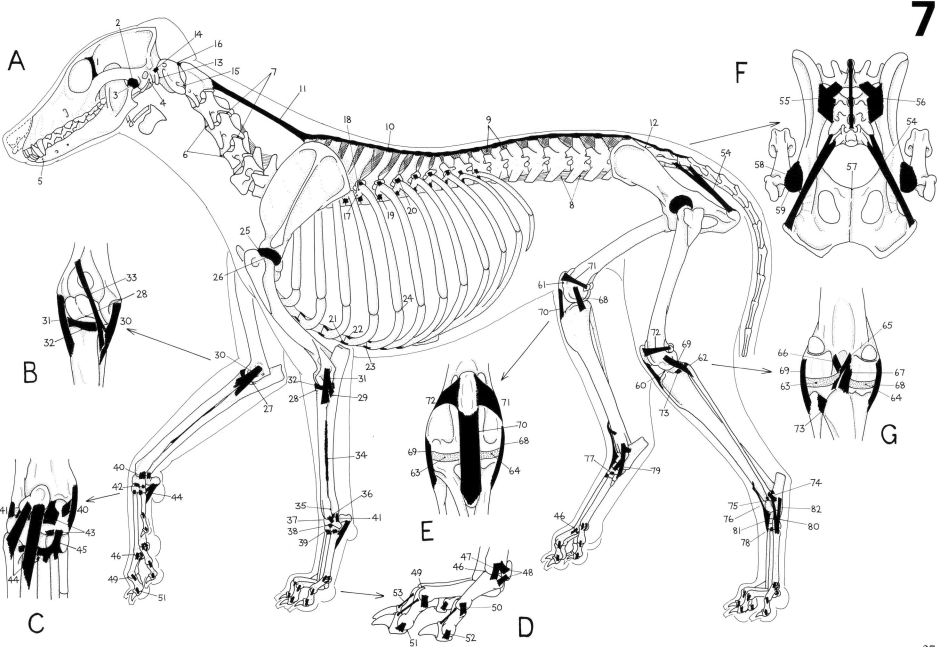

7

27

8

SUBCUTANEOUS STRUCTURES OF THE DOG – CUTANEOUS MUSCLES AND NERVES

This drawing shows a dog from which the skin has been removed to expose the immediately underlying, subcutaneous tissues. In order to remove the skin this subcutaneous tissue attaching to the dermis would have had to have been cut through and shredded. It is highly likely that during such a procedure many subcutaneous structures shown here in detail would have been removed along with the skin. This is especially true of cutaneous muscles which move the skin and cutaneous nerves which supply sensation to it; both components will be damaged to a considerable extent. Since the drawing depicts these structures intact it is to a considerable extent diagrammatic and stylized. This observation concerning the diagrammatic nature can be levelled at most of the drawings in the book. Nevertheless, the object in all has been to bring out the essential points, be they nerves, muscles, bones, etc, which will contribute to an understanding of the gross structure of a dog, rather than to adhere rigidly to pictorial (photographic) accuracy.

Subcutaneous tissue or **superficial fascia**, attaching skin to underlying structures is a loose connective tissue rich in elastic fibres and usually containing variable quantities of fat. It varies considerably in amount, just as the thickness of the skin varies. Where it is abundant, as over the neck and back, the skin tends to be more loosely attached; where it is sparse, as in the limbs, the skin is more closely attached to underlying structures.

Over large areas of the body superficial fascia contains thin sheets of **cutaneous muscle** which are able to produce limited skin movements. These muscle sheets must therefore be attached to the dermis of the skin and to the skeleton, although skeletal attachments are few. The largest of these muscles is associated with the skin of the trunk (**cutaneous trunci**) covering much of the back, loins, chest and flanks. It extends downwards and forwards from the rump converging into the axilla (armpit) where it is attached to the pectoral muscles. On vigorous contraction it shakes the skin clearing the coat of dust or moisture or any other irritants. Such an action is important in many large mammals, horses for instance, but does not have much importance in a dog which can shake its whole body vigorously or can lie down and scratch itself with its hindpaws or groom itself with its teeth and tongue quite adequately. Nevertheless, rapid cutaneous muscle contractions are responsible for the shivering movement of the skin promoting a flow of warm blood to the body surface which is important when a dog is cold since it raises the local temperature in its skin.

Additional muscle strands from the cutaneous trunci extend back on the belly into the sheath (see also fig 18B). This **preputial muscle** keeps the sheath in position as a protective covering over the sensitive glans penis. It maintains the position of the penis in the prepuce at rest and pulls the prepuce forwards and upwards over the glans as an erection subsides. Muscle is absent from the skin over the rump and thigh although the **external anal sphincter** encircling the anal canal is a modified cutaneous muscle supplementing an internal anal sphincter, the two constituting the main components of an anal diaphragm.

The cutaneous muscle of the neck (cutaneous colli) although sparse on the underside, produces a well developed sheet the **platysma muscle** which extends downwards and forwards from the back of the neck, beneath the ear and onto the cheek as far forwards as the corner of the mouth. Cutaneous muscle reaches its greatest development over the head where it is subdivided into several parts as the **muscles of facial expression** (see also fig 9B). These are basically associated with movable components of the face arranged around openings in the skin of the head. A sphincter (closure) muscle surrounds the mouth (*orbicularis oris muscle* of the lips) and another surrounds the eye (*orbicularis oculi muscle* of the eyelids); dilator (opener) muscles such as the *levator nasolabialis* are associated with the lateral wing of the nostril. Nostril dilators acting in conjunction with the maxillary incisive muscle which raises the upper lip, will bare the teeth in a snarl. Additional movements of the eyelids are brought about through a *medial levator* and a *lateral retractor muscle*, which produce the wrinkling of the forehead when a dog frowns. The external ear also has several muscles attached to the auricular cartilage which is the substance of the pinna. In combination *auricular muscles* can produce the varied movements of the ear noticeable in any reasonably alert animal.

A most important facial muscle is the *buccinator* lying in the lips and cheeks. It extends between upper and lower jaws but is only exposed after removal of the more superficially placed platysma and orbicularis oris muscles (see fig 9A). The cheek forms the outer boundary to the oral vestibule whose inner boundary is the teeth and gums. The buccinator assists in chewing by pushing food back into the mouth cavity proper from the oral vestibule across the biting surfaces of the teeth. Chewing, however, is not as important in a dog (a carnivore) as it is in animals with a diet containing more vegetable matter (herbivores). Hence the cheek, and so the buccinator muscle, is not as prominent as in many mammals.

Since the skin is the boundary between a dog and its surroundings it contains many sensory receptors. Sensations arising from such stimuli as touch, pain and temperature are transmitted from these sensory receptors in the skin in **cutaneous nerves**. The initial parts of such nerves will therefore run within the superficial fascia and many must penetrate cutaneous muscles in their course away from the skin towards the central nervous system embedded more deeply in the body. The drawing attempts to show some of the ramifications of these cutaneous nerves where they lie in a subcutaneous position. As has already been suggested, in this respect the drawing is somewhat artificial since cutaneous nerve ramifications are widespread, ultimately delicate and would most certainly have been removed almost completely on skinning the dog.

Should damage occur to a nerve supplying an area of skin, then it is quite likely that the skin in that area will be desensitized. This is the basis of local anaesthesia, when the anaesthetic agent temporarily blocks the transmission of impulses from the sensory nerves numbing the area.

Blood vessels to and from the skin run for some distance within the loose meshes of superficial fascia where they are to some extent protected and escape the injuries they might suffer were they shorter and more direct and therefore more rigidly fixed in position. The smallest vessels are minute and join up extensively to produce plexuses – networks of vessels which are not shown in the drawing because of their small size and complexity. However, some prominent veins run in superficial positions within subcutaneous tissue, notably the **cephalic vein** of the frontleg and **saphenous vein** of the hind, both of which are illustrated. These superficial vessels are of some importance since they are readily available for a veterinary surgeon to inject into or to withdraw blood samples from. In the everyday life of a dog the importance of such vessels lies in their being alternative return channels from the paw should deeper vessels be inadequate.

Blood flow in veins is slower than in arteries and depends to a considerable extent on the activities of neighbouring muscles – intermittent muscle contraction compresses veins and moves blood along. Valves present inside veins are specially designed and positioned to confine this movement to a one-way flow towards the heart. Any tendency for blood to flow in the opposite direction, away from the heart, is prevented by these valves flapping out from the walls and blocking the vein. Continuous muscle activity, on the other hand, may severely restrict blood flow in veins by permanent compression. In ourselves, restriction of the deep venous return flow from our legs, occasioned by prolonged sustained contraction as in standing still for long periods, will divert blood to the superficial saphenous veins. Since these superficial vessels run in fairly loose connective tissue providing them with little support they may become distended because of being persistently loaded with blood beyond their usual capacity. Such situations recurring over extended periods of time may produce pathological dilatation or varicosity of the saphenous veins, a price many of us must pay for our upright posture.

In certain parts of the body we have already seen that bone lies subcutaneously separated from the skin only by the superficial fascia, which may indeed be very thin. Should considerable friction or pressure occur at such a site, as for example over the point of the elbow or hock, then the fascia may be modified and a small sac of connective tissue (**subcutaneous bursa**) will form. These fibrous sacs contain small quantities of a cushioning fluid (identical to that found in synovial joints) sufficient to allow the internal surfaces to glide smoothly over one another. Such bursae help to reduce friction and pressure by acting as cushions. With continued pressure at these points throughout life it is likely that subcutaneous bursae will progressively become replaced by masses of fibrous tissue and the overlying skin will become more and more thickened and callosed. An examination of the hocks or elbows of an aged dog especially of the larger breeds might quite well reveal such thickened patches.

In similar situations of potential damage tendons may also have their passage smoothed out and cushioned by the interposition of a **synovial bursa** between themselves and bone. Synovial bursae associated with muscle tendons are found in a number of positions in the body as normal anatomical structures. It is possible, however, that a bursa may develop in response to unusual friction or pressure.

Superficial fascia merges imperceptibly with a more clearly defined deeper lying layer of more dense fascia which closely invests the underlying body muscles. This **deep fascia** is composed of very dense, fibrous connective tissue, and in many areas forms thick glistening sheets of tissue. It is particularly well developed in the limbs where it may be cut through and in some cases peeled away from the surface of underlying muscle. Numerous projections from it extend inwards between muscles separating them and attaching onto the fibrous periosteum of bone. Therefore, quite distinct fascial sheets separate muscles from each other facilitating their action, but also providing routes for blood vessels, nerves and lymphatics, as well as being a place in which fat can be stored. Fascial planes might also represent pathways along which pus could spread from an infected area. On the other hand fascial layers do represent quite an effective check to the spread of infective material through the body. Thus despite fascia appearing at first sight to be a rather uninteresting tissue, it is complex and widespread.

In the lower parts of the limbs (forearm and crus) where deep fascia reaches its most prominent development it forms a tight 'sleeve' around muscles restricting them and directing their action along specific lines. This restraining sleeve also means that under normal circumstances blood and tissue fluids generally are prevented from accumulating. At the carpus and tarsus long tendons from flexor and extensor muscles situated higher up in the forearm and shank, pass down into the paw. They are held in position against the surface of carpal and tarsal bones by thickened bands of deep fascia (**retinacula**) which cover them and blend with the fibrous component of periosteum on the bone surface. In such a situation a tendon might be subjected to friction over its entire surface because it is running inside a tunnel composed of deep fascia and bone. A synovial bursa would only be of limited value to a tendon in this case and is replaced by a **synovial tendon sheath**. This may be likened to a bursa which has become enlarged and completely wrapped around the tendon to form a tube; the outer layer is in contact with the tunnel walls, the inner layer with the tendon. When the tendon moves the two layers slide over one another lubricated by a small quantity of synovial fluid. Synovial tendon sheaths are an important adjunct to many tendons as they cross the carpus and tarsus and pass along the digits (fig 11).

Since many areas of deep fascia are

especially thick and well formed, they may provide attachment areas for muscles themselves, just like skeletal bones. The thoracolumbar and gluteal fascia, both shown in this drawing, can be included in this category, the former providing attachments for muscles of the abdominal wall. In the flanks and belly, however, deep fascia is not so well developed as it is on the limbs and is not as easily separated from underlying muscles. The relatively poor development of deep trunk fascia presumably allows for the degree of expansion required from this area. Nevertheless it must be borne in mind that the abdominal wall in particular is supporting the weight of the guts through its muscles. In this action it receives assistance from its deep fascia.

Bony landmarks
1 Spine of scapula. 2 Olecranon process of ulna (point of elbow). 3 Accessory carpal bone. 4 Sacral tuberosity of ilium. 5 Calcaneal tuberosity (point of hock).

Muscles
6 Superficial sphincter muscle of neck (located in superficial fascia of neck). 7 Platysma muscle. 8 Cutaneous muscle of trunk (located in superficial fascia of trunk). 9 Preputial muscle of sheath. 10 Deep sphincter muscle of neck and head (intermediate part). 11 Frontal muscle. 12 Orbicularis oculi muscle of eyelids (sphincter of palpebral fissure). 13 Orbicularis oris muscle of lips (sphincter of mouth). 14 Levator muscle of nostril wing and upper lip. 15 Rostral auricular muscles. 16 Brachiocephalic muscle. 17 Trapezius muscle. 18 Deltoid muscle. 19 Lateral head of triceps muscle. 20 Biceps femoris muscle. 21 Common calcaneal tendon.

Fascial layers
22 Superficial fascia of rump and tail. 23 Superficial fascia of shoulder and arm. 24 Deep fascia of forearm (antebrachial fascia forming a close fitting sleeve around forearm muscles). 25 Deep fascia of carpus (producing flexor and extensor retinacula). 26 Deep fascia of thigh (lateral femoral fascia – providing attachment for biceps femoris muscle and having its own tensor muscle). 27 Deep fascia of shank (crural fascia – closely investing muscles of crus). 28 Deep fascia of tarsus (producing flexor and extensor retinacula of tarsus).

Blood vessels and lymph nodes
29 Facial vein. 30 Linguofacial vein. 31 Maxillary vein. 32 External jugular vein. 33 Dorsal common digital veins. 34 Accessory cephalic vein. 35 Cephalic vein. 36 Axillobrachial vein. 37 Omobrachial vein. 38 Lateral saphenous vein. 39 Medial saphenous vein. 40 Mandibular lymph nodes.

Cutaneous branches of cranial nerves
41–51 Trigeminal nerve (cranial nerve 5). 41–43 Ophthalmic branch of trigeminal nerve (smallest of 3 trigeminal subdivisions – sensory nerve of orbit, skin of dorsum of muzzle and some of mucous membrane of nasal cavity and paranasal sinuses). 41 Frontal (supraorbital) nerve. 42 Infratrochlear nerve (from nasociliary branch of ophthalmic). 43 External nasal nerve (from ethmoidal branch of ophthalmic). 44–46 Maxillary branch of trigeminal nerve (largest of 3 trigeminal subdivisions – sensory nerve of cheek, side of nose, muzzle, maxillary recess, palate, teeth and gums of upper jaw). 44 Zygomaticotemporal nerve. 45 Zygomaticofacial nerve. 46 Infraorbital nerve. 47–51 Mandibular branch of trigeminal nerve (mixed motor and sensory nerve – sensory components to

pinna, cheek, lower lip, teeth and gums of lower jaw). 47–48 Auriculotemporal nerve. 47 Rostral auricular nerve (from auriculotemporal). 48 Transverse facial nerve (from auriculotemporal). 49 Buccal nerve. 50 Mental nerves. 51 Mylohyoid nerve.

Cutaneous branches from dorsal rami of spinal nerves
52 Cutaneous branches of dorsal rami of cervical nerves 2–7. 53 Cutaneous branches of dorsal rami of thoracic nerves 2–13. 54 Cutaneous branches of dorsal rami of lumbar nerves 1–6 (cranial clunial nerves). 55 Cutaneous branches of dorsal rami of sacral nerves 1–3 (middle clunial nerves). 56 Cutaneous branches of dorsal rami of caudal nerves 1–5.

Cutaneous branches from ventral rami of cervical nerves
57–58 Ventral ramus of cervical nerve 2. 57 Great auricular nerve. 58 Transverse cervical nerve. 59 Lateral cutaneous branches of ventral rami of cervical nerves 3–6 (supraclavicular nerves). 60 Lateral thoracic nerve (from ventral rami of cervical nerve 8 and thoracic nerve 1).

Cutaneous branches from ventral rami of thoracic nerves (intercostal nerves)
61 Lateral cutaneous branch of intercostal nerve 2 (intercostobrachial nerve). 62 Lateral cutaneous branch of intercostal nerve 3 (intercostobrachial nerve). 63 Lateral cutaneous branches of intercostal nerves 4–12. 64 Lateral cutaneous branch of costoabdominal nerve (thoracic nerve 13). 65 Ventral cutaneous branches of intercostal nerves 3–10.

Cutaneous branches from ventral rami of lumbar nerves
66 Lateral cutaneous branches of ventral rami

of lumbar nerves 1–3 (cranial iliohypogastric, caudal iliohypogastric and ilioinguinal nerves). 67 Genitofemoral nerve (from ventral ramus of lumbar nerve 4). 68 Lateral cutaneous femoral nerve (bulk of ventral ramus of lumbar nerve 4).

Cutaneous branches from ventral rami of sacral and caudal nerves
69 Caudal cutaneous femoral nerve (from ventral rami of sacral nerves 1 and 2 and giving origin to perineal branches and caudal clunial nerves). (Ventral rami of sacral nerves 1–3 also give rise to a pudendal nerve and a caudal scrotal nerve). 70 Cutaneous branches from ventral rami of caudal nerves 1–5.

Cutaneous nerves from brachial plexus to forelimb
71 Cranial lateral cutaneous brachial nerve (from axillary nerve). 72 Superficial radial nerve (from radial nerve). 73 Cranial cutaneous antebrachial nerve (branches from medial and lateral elements of superficial radial nerve). 74 Lateral cutaneous antebrachial nerve (from lateral branch of superficial radial nerve). 75 Caudal cutaneous antebrachial nerve (from ulnar nerve). 76 Medial cutaneous antebrachial nerve (from musculocutaneous nerve). 77 Dorsal branch of ulnar nerve.

Cutaneous nerves from lumbosacral plexus to hindlimb
78 Saphenous nerve (from femoral nerve). 79 Lateral cutaneous sural nerve (from common peroneal [fibular] nerve). 80 Cutaneous branch from superficial peroneal [fibular] nerve. 81 Caudal cutaneous sural nerve (from tibial nerve). 82 Cutaneous branches of medial and lateral plantar nerves (from tibial nerve).

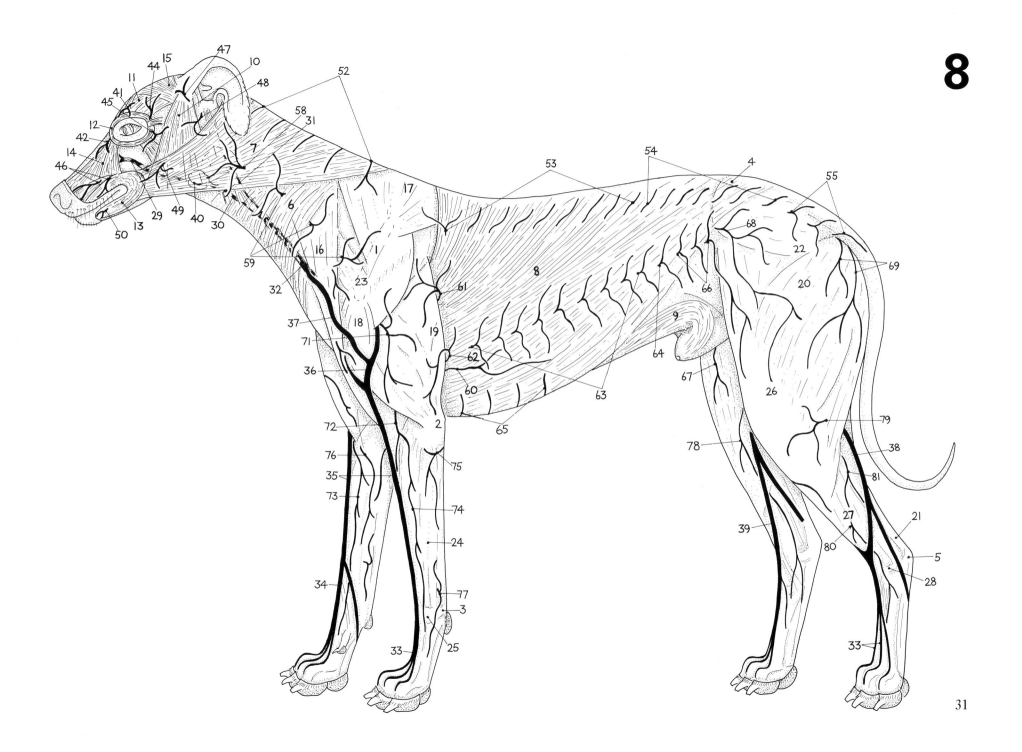

8

31

9

SUPERFICIAL MUSCLES OF THE DOG

Muscle is the body tissue concerned with movement being made up of cells able to contract (shorten in length) and so move structures to which they are attached. Coordinated contractions in numerous muscles will result in movement of the body overall, but will also move the mouth, teeth, jaws and tongue in such actions as obtaining food and grooming; move the head, eyes, ears and nose in appraisal of the surroundings; move food through the guts, air into and out of the lungs, semen through the reproductive tract, and blood through the vascular system; move the hair in response to cold and fear, and the skin in response to cold and irritation. Such an array of activities will need variation in both speed and force of muscle contraction and, since some of the activities are continuous throughout life, some contraction will need to be independent of the voluntary control (will) of a dog. Different types of muscle tissue can therefore be recognised.

Firstly, there is **skeletal muscle** displaying powerful but short-lived contractions and making up organs in their own right under a dog's voluntary control. Skeletal muscles form the flesh or meat of the body and are used by a dog to adjust to its surroundings since they attach to and move skeletal structures producing gross body movements,

movements of eyes, ears and nose, and opening and closing of orifices such as the anus.

Secondly, **visceral muscle** is weaker but displays longer-lasting contractions and is found only as a contributory component of various internal organs that are otherwise non-muscular; eg. the muscles of the gut wall associated with the automatic movements of food churning and food passage. In blood vessel walls it controls the vessel diameter and consequently the volume of blood flowing through it. In this way blood supply to any part of the body can be regulated. It also operates hairs in the skin, raising or lowering them, and produces secretions from glands such as sweat. Visceral muscle is automatic (involuntary) and its contraction is regulated through the autonomic nervous system. In the mouth and pharynx (throat) the visceral musculature is modified so that in microscopic structure and in function it resembles skeletal voluntary muscle. This enables the jaw and pharyngeal muscles to perform the more complex, faster and more delicately balanced feeding and respiratory functions encountered in the head region. Like skeletal muscles they are now under the voluntary control of a dog and are illustrated in the accompanying drawings.

Thirdly, **cardiac muscle** displays strong, rhythmic and incessant contractions and is confined to the heart. In microscopic structure cardiac muscle cells have some similarity to both skeletal and visceral muscle although it is involuntary, its contractions being inherent and rhythmic requiring no nerve stimulus. But, like visceral muscle, its

rate of contraction is regulated by the autonomic nervous system.

Visceral muscle, since it does not form organs in their own right merely being one component of an organ, can have little further categorization apart from topographical; eg. muscles of the gut tube, muscles of the reproductive tract, etc. Skeletal muscles, however, can be extensively subdivided into separate categories. This subdivision may be based on topography like visceral muscle, so that we can recognize head muscles, neck muscles, forelimb muscles, and so on. But just as with the skeleton where we suggested that axial and appendicular components were easily recognized, we may also recognize these two broad groupings of muscles:

Axial musculature of the head, neck, trunk and tail associated with the skull, vertebral column and ribcage.

Appendicular muscle of the limbs associated with the skeleton of the limbs and girdles.

Yet another categorization could be made with reference to functional activity; eg. extensor muscles of the shoulder joint, flexor muscles of the hock joint, adductor muscles of the hip joint, and so on. This functional categorization is undoubtedly useful since it conveys information about position and group action of muscles and will be used later on. Complementing this functional categorization is one based on the nerve supply to muscles. The motor nerve innervation to muscles is organized in terms of these same functional muscle groupings as we will also consider later (fig 34). Knowing the function and motor innervation of a muscle might enable you to predict

impairment to movement following nerve damage.

This and the following nine or ten drawings of musculature concentrate on skeletal muscles, the identifiable body muscles which are organs in their own right forming the flesh or meat of the body. This first drawing shows what a dog might look like after the skin and its associated cutaneous muscles have been removed. In subsequent drawings removal of superficial musculature shows the more deeply lying muscles, and progressive removal of these ultimately reveals the contents of thoracic, abdominal and pelvic cavities. The inset drawing at the left of the page (**B**) still shows the facial musculature in some detail.

A skeletal muscle consists of numerous contractile muscle cells joined together by connective tissue. Such connective tissue therefore forms a significant component of any muscle and, since muscle cells and the cells in connective tissue are living structures requiring both blood and nerve supply, blood vessels and nerves also ramify throughout muscle in its connective tissue. A muscle also needs to be attached in the body, so its internal connective tissue extends beyond the boundaries of its contractile muscle cells (the belly of the muscle) as tendon. The connective tissue will concentrate and transfer the power developed during contraction through the tendon to its attachments. The two gross muscle components that the drawings therefore show are the **belly** and **tendon**. Both demonstrate enormous variation in shape and disposition as you will readily appreciate from the drawings.

A muscle belly may be large (eg. the biceps femoris component of the hamstring group of muscles in the thigh) or small (eg. the muscles moving the eyeball in its socket): it may be flattened into a sheet (eg. the abdominal wall muscles in the flanks and belly) or stretched out into a long flattened strap (eg. the sartorius muscle of the thigh): it may be irregularly shaped to fit into a particular area of the body (eg. the deltoid muscle lateral to the shoulder joint) or it may be regular and possibly fusiform in shape (e.g. the digital flexor and extensor muscles of the forearm and crus): it may be a single structure as most body muscles are or it may be arranged as a structure broken up into numerous segments (eg. the intercostal muscles between the ribs).

A typical tendon appears as a cord or band-like structure like those in the lower parts of the limbs, especially in the carpal and tarsal regions and below. Tendons, nevertheless, assume different forms according to the position, size and shape of the muscle belly of which they are a continuation. Many, for instance, are flattened sheets (aponeuroses) associated with flat sheets of muscle like those of the abdominal wall. Tendons need not necessarily attach directly to bone although ultimately some skeletal attachments occur. Thus the abdominal muscle tendons appear to be attached to a fibrous area in the ventral midline of the belly, the linea alba, and to the fascia of the thoracolumbar region in the back and loin. Both of these structures, nevertheless, do themselves have skeletal attachments. Tendons may even apparently be absent as in, for example, sphincter muscles like the external anal sphincter which completely encircle an opening without having any obvious tendon.

Despite some exceptions, it is reasonable to suggest that most skeletal muscles produce actions at particular joints over which they pass, and are demonstrably attached at either end, these attachments being termed the origin and insertion. In the performance of a specific activity that attachment of a muscle which remains more stationary is its **origin**; its **insertion** is the more mobile end at which the moving force is applied. Only rarely, however, does a muscle have a single action so that the assignment of origin and insertion may be somewhat arbitrary. Also, in the limbs especially, both points of attachment may be movable, thus the more proximal attachment is generally taken to be its origin.

The angle at which a muscle through its tendon of insertion attaches to a bone, and the distance of that attachment from the joint being acted upon, will both be factors which influence the potential range of movement of that joint. If a muscle is inserted on a bone at right angles to it then its entire contractile activity will be used to move the joint that it crosses: if a muscle is inserted at a very acute angle to a bone practically its entire force when it contracts will tend to stabilize and hold a joint together rather than move it. Most muscles will lie somewhere between these two theoretical extremes of activity, contraction being made up of a **moving component** producing angular displacement between the bones making up the joint, and a **stabilizing component** tending to hold together the bones making up the joint.

It would appear that the ability of muscles to produce movement must be coupled with a capacity to be able to prevent it. In a dog standing up normally numerous skeletal muscles are therefore concerned solely with preventing movement by maintaining joint stability. Similarly, muscle sphincters of the guts temporarily stop the passage of its contents.

Finally, in this consideration of muscle activity there are two particularly important attributes that we have so far not mentioned. Firstly, it is rare for a muscle to produce an action on its own — it normally functions as one component of a group of muscles. Secondly, muscles can only produce movement when they contract — when a muscle relaxes movement does not result unless the part which has been moved is acted upon by the contraction of another group of muscles (or is affected by the pull of gravity). These two attributes suggest that muscles are almost always arranged in opposing (antagonistic) groups performing opposite actions on any given joint. Cooperation between opposing muscle groups is an especially important function of overall coordinated muscle activity.

In any movement, particularly should it be slow and deliberate, the antagonist (opposer) is extremely important acting at the same time as its agonist (prime mover) in order to steady the joint being moved. As an agonist contracts to produce a movement its antagonist is undergoing a gradual, 'active' relaxation to an equivalent degree, the balance between the two smooths and controls the movement. Also, in rapid movements the ligaments reinforcing most joints are subjected to sudden and considerable strains and might be severely damaged were the rapid movements not checked in some way. This is particularly the case in the limbs because of the considerable momentum that they build up during their movement. To prevent such injury antagonistic muscle groups contract to check movement. As they contract the agonists are relaxing, if indeed they have not done so already, allowing the momentum built up to complete the movement.

Skeletal muscles are therefore organs which can move body parts. At the same time many of these same muscles must also cooperate together to prevent movement of body parts. In the limbs, for instance, the skeleton on its own is a flexible structure united by ligaments of fibrous connective tissue. Obviously each individual bone is rigid and thus resistant to squashing and stretching, but the limbs do not collapse because muscles are acting across the joints between bones. The same muscles that act antagonistically to move joints and bring about body movement must also cooperate to stabilize those same joints and convert the limb into a rigid support during standing. For example, in a normal standing position the elbow is held in an extended condition by isometric contraction in the triceps muscle attached to the olecranon process of the ulna (point of the elbow); ie. tension is increased in the muscle without bringing about shortening, the resistance against which the muscle is acting is the animal's weight (pull of gravity) tending to flex the elbow. In elbow joint movement the triceps contracts concentrically, the muscle shortening to extend the elbow. Stabilization (fixation) of many other joints depends upon the cooperation between antagonistic muscle groups to a greater extent; ie. equal isometric contraction in both groups effectively cancelling out any tendency for either to produce joint movement.

Bones and ligaments
1 Zygomatic (supraorbital) process of frontal bone. 2 Nasal bone. 3 Zygomatic arch (bridge of bone connecting face and cranium below eye). 4 Spine of scapula. 5 Caudal angle of scapula. 6 Medial (flexor) epicondyle of humerus. 7 Lateral (extensor) epicondyle of humerus. 8 Body (shaft) of radius. 9 Styloid process of radius. 10 Olecranon process of ulna (point of elbow). 11 Styloid process of ulna. 12 Accessory carpal bone. 13 Crest of ilium. 14 Position of patella (in tendon of quadriceps femoris muscle). 15 Body (shaft) of tibia. 16 Medial malleolus of tibia. 17 Lateral malleolus of fibula. 18 Calcaneal tuberosity (point of hock). 19 Rib 5. 20 Rib 13 (last or floating rib).

Muscles of head
21 Platysma muscle. 22–37 Facial musculature (muscles of facial expression). 22 Intermediate sphincter muscle of head and neck. 23 Frontal muscle. 24 Orbicularis oculi muscle of eyelids (sphincter muscle of palpebral fissure). 25 Orbicularis oris muscle of lips (sphincter muscle of mouth). 26 Levator muscle of medial (inner) angle of eye. 27 Retractor muscle of lateral (outer) angle of eye. 28 Levator muscle of upper lip. 29 Levator nasolabialis muscle (levator muscle of nostril wing and upper lip). 30 Zygomatic muscle (retractor muscle of angle of mouth). 31 Pal-pebral part of sphincter muscle of head (malar muscle). 32–33 Buccinator muscle (forming basis of cheek). 32 Buccal part of buccinator muscle. 33 Molar part of buccinator muscle. 34 Zygomaticoauricular muscle. 35 Parotidoauricular muscle. 36 Interscutular muscle. 37 Scutuloauricular muscle. 38–40 Jaw muscles. 38 Temporal muscle (passing from temporal fossa of cranium onto coronoid process of mandible). 39 Masseter muscle (passing from zygomatic arch onto masseteric fossa of mandible). 40 Digastric muscle (passing from jugular process of occiput onto body of mandible). 41 Geniohyoid muscle (passing from mandible close to symphysis back to basihyoid bone).

Muscles of neck, trunk and tail
42 Sternohyoid muscle. 43 Sternocephalic muscle. 44 External abdominal oblique muscle (originating from ribs and from thoracolumbar fascia). 45 Tendon of insertion (aponeurosis) of external abdominal oblique muscle (meeting fellow of opposite side in midventral fibrous linea alba). 46 Tail depressor muscles (ventral sacrocaudals). 47 Tail levator muscles (dorsal sacrocaudals). 48 Lateral flexor muscles of tail (intertransverse caudals). 49 Coccygeus muscle (component of pelvic diaphragm extending between ischiatic spine and caudal vertebrae and forming medial boundary of ischiorectal fossa).

Muscles of forelimb
50–52 Brachiocephalic muscle (important limb protractor muscle). 50 Cleidobrachial part of brachiocephalic muscle. 51 Cleidocervical part of brachiocephalic muscle. 52 Clavicular tendon (fibrous intersection representing remains of clavicle situated in brachiocephalic muscle). 53–54 Trapezius muscle. 53 Cervical part of trapezius muscle. 54 Thoracic part of trapezius muscle. 55 Omotransverse muscle. 56 Latissimus dorsi muscle (important limb retractor muscle). 57 Deep pectoral muscle (adductor muscle of arm and subsidiary limb retractor). 58–59 Deltoid muscle. 58 Scapular part of deltoid muscle. 59 Acromial part of deltoid muscle. 60 Supraspinous muscle. 61 Infraspinous muscle. 62 Teres major muscle. 63 Biceps brachii muscle (shoulder extensor and elbow flexor muscle). 64 Brachial muscle (elbow flexor muscle inserting into forearm with biceps). 65–67 Triceps muscle (elbow extensor muscle and main support of forelimb in normal standing posture). 65 Long head of triceps muscle (arising from caudal border of scapula and forming caudal margin of arm). 66 Lateral head of triceps muscle. 67 Medial head of triceps muscle. 68 Pronator muscle of forearm. 69–72 Flexor muscles of carpus and digits. 69 Radial carpal flexor muscle. 70 Ulnar carpal flexor muscle (forming caudal contour of forearm). 71 Superficial digital flexor muscle and tendon. 72 Radial head of deep digital flexor muscle. 73–77 Extensor muscles of carpus and digits. 73 Radial carpal extensor muscle (forming cranial contour of forearm). 74 Common digital extensor muscle and tendon. 75 Lateral digital extensor muscle and tendon. 76 Ulnar carpal extensor muscle (lateral ulnar muscle). 77 Oblique carpal extensor muscle and tendon (abductor muscle of digit 1).

Muscles of hindlimb
78–80 Rump muscles. 78 Superficial gluteal muscle. 79 Middle gluteal muscle. 80 Tensor muscle of lateral femoral fascia. 81–82 Sartorius muscle (forming cranial margin of thigh). 81 Cranial part of sartorius muscle. 82 Caudal part of sartorius muscle. 83–85 'Hamstring' muscles (extensor muscles of hip joint and important limb retractor muscles). 83 Biceps femoris muscle. 84 Semitendinosus muscle. 85 Semimembranosus muscle. 86–89 Extensor muscles of digits and flexor muscles of tarsus. 86 Cranial tibial muscle and tendon. 87 Long digital extensor muscle and tendon. 88 Lateral digital extensor muscle and tendon. 89 Long peroneal muscle and tendon. 90–93 Flexor muscles of digits and extensor muscles of tarsus. 90 Lateral head of gastrocnemius muscle. 91 Medial head of gastrocnemius muscle. 92–93 Deep digital flexor muscle of hindlimb. 92 Medial digital flexor muscle. 93 Lateral digital flexor muscle. 94 Accessory (tarsal) tendon of 'hamstring' and gracilis muscle (thickened band of deep crural fascia extending down onto point of hock). 95 Common calcaneal tendon (aggregate of structures attaching to point of hock, including Achilles' tendon, superficial digital flexor tendon, and accessory or tarsal tendon of 'hamstring' and gracilis muscles).

Fascial layers
96 Thoracolumbar fascia (deep fascial layer enclosing epaxial muscles and providing attachment for latissimus dorsi and lateral abdominal muscles). 97 Lateral femoral fascia (deep fascial layer enclosing extensor muscles of stifle joint and providing one area of attachment for biceps femoris muscle). 98 Extensor retinacula (loops of deep fascia holding extensor tendons in position at carpus and tarsus). 99 Flexor retinacula (holding flexor tendons in position at carpus and tarsus and completing carpal and tarsal canals for passage of deep digital flexor tendons).

Salivary glands and ducts
100 Parotid salivary gland (diffuse and wrapped around concha of auricular cartilage). 101 Parotid duct (crossing masseter muscle and opening into oral vestibule through cheek). 102 Mandibular salivary gland (caudomedial to angle of jaw).

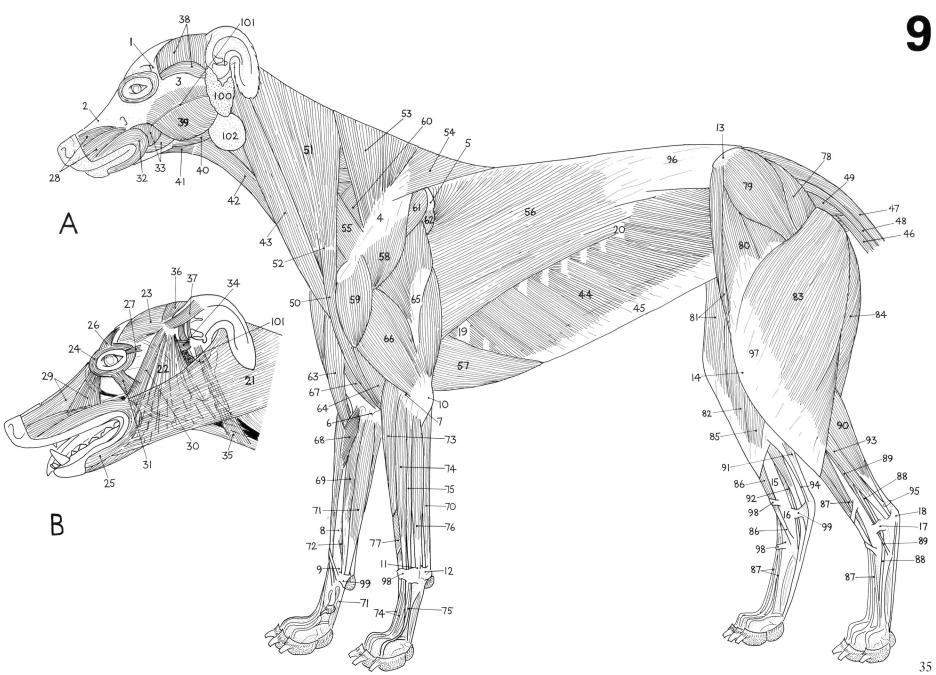

9

A

B

28

35

10

APPENDICULAR MUSCLES OF
THE DOG – FORELIMBS &
HINDLIMBS

This second muscle drawing deals with the appendicular muscles, the limbs having been removed from the body. In order to remove a forelimb from the trunk it would have been necessary to cut through a number of muscles. But, if you recall the structure of the skeleton it would not have been necessary to cut through any bones. Those muscles severed therefore connect the limb and its girdle (scapula) to the trunk and are sometimes referred to as **extrinsic muscles** of the limb distinguishing them from **intrinsic muscles** confined to the limb itself. The muscles which were cut through are the pectorals, brachiocephalic, omotransverse, trapezius, rhomboid, latissimus dorsi and ventral serrate.

The organization of muscles into antagonistic groups outlined earlier is particularly well displayed in the limbs since the more distal joints are straightforward 'hinges' in action (although not necessarily in anatomical conformation) and are operated by opposing flexor/extensor muscle groups to some extent apparent from the drawings.

In the forelimb a muscle mass located on the outer and cranial aspect of the forearm comprises essentially a pair of digital extensor muscles and a pair of carpal extensors. These arise primarily from the lateral (extensor) epicondyle at the lower end of the humerus, all four having tendons crossing the carpus, with the digital tendons continuing and inserting onto the extensor process of a third phalanx. The detailed distribution of these tendons is shown in the paw illustration later (fig 11). The antagonist group of forearm muscles is located caudally and medially and again consists essentially of a pair of carpal flexors and a pair of digital flexors. These basically arise from the medial (flexor) epicondyle of the lower end of the humerus, although the deep digital flexor and the ulnar carpal flexor have additional origins from the upper ends of both radius and ulna. The detailed distribution of the tendons in the paw can be seen in fig 11.

In both posture and movement flexors are considerably more important, especially for support of the carpal and metacarpophalangeal joints, and for sinking the claws into the ground to obtain firmer footing. Consequently the flexor muscles in the forearm are larger, more powerful muscles than the extensors. The **ulnar carpal flexor muscle** attached to the accessory carpal bone plays a particularly important role in preventing overextension and consequent collapse of the carpus during normal standing. It can always be felt as a taut cord at the back of the forearm above the accessory carpal bone whenever a dog is supporting weight on that limb. Subsidiary actions of these flexor muscles will also include quite possibly, elbow extension since they cross the extensor surface of the elbow caudal to its centre of rotation (see fig 19). However, at the elbow it is more likely that they simply assist elbow fixation during normal standing, maintaining the elbow joint in an extended condition.

In the hindlimb a muscle mass located on the outer and cranial aspect of the crus comprises basically a pair of digital extensor muscles and a pair of tarsal flexors. These arise from the lateral femoral condyle, lateral tibial condyle and the proximal lateral surface of both tibia and fibula. Distribution in the digits is essentially the same as in the forelimb and the detailed distribution of these tendons and their synovial tendon sheaths at the tarsus is illustrated in fig 11. The opposing group of tarsal and digital muscles forms the muscle mass of the calf at the rear of the crus. **Superficial** and **deep digital flexors** are present, the latter is larger and more powerful and originates from the rear surfaces of the upper ends of both tibia and fibula. The former originates on the rear surface of the lower end of the femur. In the paw the tendons of both flexors divide and insert as in the forelimb (fig 11). However, *en route* the superficial tendon expands and caps the calcaneal tuberosity (point of the hock) to which it is attached. As well as digital flexion both muscles extend the hock while the superficial component may also assist in stifle flexion. The **gastrocnemius muscle** in this caudal group originates from supracondylar tuberosities at the rear of the lower end of the femur, and its tendon of attachment, **Achilles' tendon**, inserts on the calcaneal tuberosity. Together with the tendon of the superficial digital flexor, and an accessory (tarsal) tendon from the hamstring and gracilis muscles, Achilles' tendon forms a **common cal-caneal tendon** extending the tarsal joint. The gastrocnemius, along with the superficial digital flexor, also has the potential to flex the stifle, but may play a role in stifle stabilization when acting in a cooperative antagonism with such a muscle as the long digital extensor.

From the point of view of limb stability the critical joints are the elbow and stifle and inability to operate either in the correct manner might mean that a dog is unable to accept weight on that limb so that it will stand and move abnormally. Despite their complexities of structure both joints function as hinges with opposing antagonistic groups of flexor and extensor muscles. At the elbow the large and strong **triceps muscle group** at the rear of the arm inserts on the point of the elbow and is concerned with supporting body weight against the pull of gravity by keeping the elbow extended. Of the four parts of the triceps three originate from the humerus and are therefore specifically elbow extensors, the fourth part, the long head has an origin on the rear of the shoulder blade so it will also function in shoulder flexion. At the stifle joint the **quadriceps femoris muscle group** at the front of the thigh is the equivalent extensor muscle mass maintaining stifle extension against the pull of gravity. It is composed of three vastus muscles arising from the femoral shaft, and a rectus femoris component originating from the hip bone in front of the hip joint. All four parts insert onto the tibial tuberosity through the **patellar tendon** which contains the sesamoid patella. The rectus component also flexes the hip since it crosses its front surface.

Flexion of elbow and stifle joints is of lesser importance than their extension occurring when the limb is being protracted (raised from the ground and swung forwards). At the elbow the **biceps** and **brachialis**, the former arising from the supraglenoid tubercle of the scapula, the latter from the proximal caudal humeral surface, both insert onto the radial and ulnar tuberosities of the forearm bones. It is evident from these attachments that the brachialis can only function as an elbow flexor while the biceps can also extend the shoulder. However, this additional activity of the biceps is more likely to contribute to shoulder stabilization, holding the shoulder joint components in apposition, rather than being a predominantly moving action. At the stifle flexion is brought about as we have suggested by subsidiary activity of the tarsal extensors such as the gastrocnemius and superficial digital flexor crossing its rear surface. It will also be due in some measure to hamstring activity since the lower insertion of the hamstrings through crural fascia is onto the tibia.

The *hamstrings* (**biceps femoris, semitendinosus** and **semimembranosus**) originate on or close to the ischiatic tuberosity (point of the buttock) and are the main hip extensor muscles generating much of the power thrust in locomotion. In this activity their insertions through the lateral femoral fascia onto the patella (biceps femoris) and onto the medial femoral epicondyle (cranial belly of semimembranosus) will also extend the stifle adding further

forward thrust. At the same time as extending stifle and hip joints, the hamstrings through their **accessory (tarsal) tendons** attaching to the calcaneal tuberosity (point of the hock) will also extend the hock. I think you can appreciate from this how important the hamstrings are in movement.

The limb joints considered so far have all been hinges, functionally if not anatomically. **Shoulder** and **hip joints**, however, are ball-and-socket joints capable of wider ranges of movement. Consequently muscles surrounding these joints will tend to be more complex in their arrangement. The hip is a pivot about which movements of the hindlimb on the trunk (or *vice versa*) occur. Although ball-and-socket, hip movement during normal locomotion is predominantly restricted to a fore-and-aft longitudinal plane, flexing and extending, with some limited degrees of abduction, adduction and rotation.

It was suggested above how hamstring muscle contraction extending the hip will effectively produce backward movement (retraction) of the entire limb as a lever because of their actions on stifle and hock as well. With the paw on the ground limb retraction will mean forward movement of the body the essential thrust in normal locomotion. Hip extension is also assisted by the **gluteal muscles** of the rump running from the hip bone onto the greater femoral trochanter. Hip flexion occurs when the limb is picked up and swung forwards (protracted) prior to placement of the paw on the ground at the beginning of the next step. It is brought about by **iliopsoas muscle** activity, but also by

such muscles as the **sartorius** forming the cranial margin of the thigh.

Abduction (outward movement of the hindlimb) at the hip occurs to some extent during every step a dog takes due simply to body weight being accepted on the limbs tending to splay them apart. It is also encountered in many correcting movements to maintain balance. However, a dog cocking his leg at a lamp post is displaying the extreme of hip abduction and is using his gluteal muscles to great effect. It is evident therefore that adduction (inward movement) of the leg at the hip joint must occur all the time that a limb is supporting weight so that limb abduction is prevented. The main muscles of adduction are the **adductors** and **gracilis** on the inside of the thigh running from the hip bone to the femur. The gracilis, like the hamstrings, is a complex muscle and extends into the lower part of the leg to attach to the tibia and, through an **accessory (tarsal) tendon**, to the point of the hock. Consequently it contributes significantly to limb retraction in the power thrust from the hindlimb.

Rotation of the femur at the hip joint can occur to some extent. During movement outward (lateral) rotation of a thigh occurs as it is being protracted (swung forwards) carrying the stifle joint slightly outwards so that it will not hit the belly or flank. Such rotation is due to a selection of small muscles caudal to the hip including the **internal obturator** and **gemelli muscles**. Medial (inward) rotation occurs in the supporting limb in the course of a stride as it is being retracted in relation to the body. Inward rotation is produced in the main by parts of the gluteal muscles.

The shoulder region has a complex muscular arrangement since in addition to the shoulder joint itself, a ball-and-socket synovial joint equivalent to the hip, there is also a muscular joint between scapula and trunk, the shoulder girdle articulation. This arrangement suggests that *the range of movements apparently exhibited by the true shoulder joint really consist of a combination of limited shoulder joint movements together with more extensive movements of the whole limb on its muscular attachments to the trunk.* This is especially true for limb adduction and abduction. Freeing the forelimbs from a bony connection with the trunk also means that they can act more efficiently as shock absorbers during movement; the shock encountered on landing being absorbed by those muscles and tendons holding the scapula in place. In addition the shoulder girdle lies in close relation to the ribcage which is involved in breathing movements. A more rigid attachment of girdle to trunk might well hamper breathing by restricting rib movements.

A greater ability to move its forelimb in relation to its trunk will also be useful for a dog moving fast and increasing the length of its stride. The scapula now moves as a unit with the limb, sliding backwards and forwards over the chest wall. This increased freedom of the frontleg also means that it can be abducted and adducted to greater degrees, adding to stability and balance during manoeuvring movements such as cornering.

At the shoulder joint itself movements occur predominantly in a fore-and-aft plane, flexing and extending, despite its ball-and-socket nature. Probably the main shoulder extensor, and protractor of the forelimb, is the **brachiocephalic muscle**. More minor extensors include the **supraspinatus** and **subscapular muscles**. These two muscles, however, are very important shoulder stabilizers with their strong tendons of insertion lying close to the shoulder joint itself and playing the part of lateral and medial 'ligaments' which restrict adduction and abduction confining the joint to a single plane of movement. The counterpart to the brachiocephalic on the rear of the shoulder is the **latissimus dorsi**. Although it will flex the shoulder to a certain extent its major action is to pull the whole limb backwards in relation to the trunk (retraction). Additional shoulder flexors include the **teres muscles**, the **deltoid** and the **infraspinatus**. However, this last muscle is best viewed as a further ligament muscle which stabilizes the shoulder joint.

Abduction and adduction at the shoulder joint refer to inward and outward movement of the humerus in relation to the scapula. The deltoid and infraspinatus muscles could abduct the humerus, the subscapular could adduct it. However, both movements appear to be of negligible occurrence so that synchronized contraction within these three muscles effectively cancels out either action stabilizing the joint. Any apparent abducting/adducting movements at the shoulder are most likely to be movements of the whole limb on its muscular attachments to the trunk. Finally, rotation of the humerus relative to the scapula might be brought about by the subscapular, infraspinatus and teres minor muscles. However, rotation does not seem to occur these three muscles presumably cooperating to prevent it.

Bones and joints of forelimb
1 Dorsal (vertebral) border of scapula. **2** Caudal angle of scapula. **3** Spine of scapula. **4** Acromion process of scapula. **5** Greater tubercle of humerus (point of shoulder). **6** Deltoid tuberosity of humerus. **7** Humeral condyle. **8** Medial (flexor) epicondyle of humerus. **9** Lateral (extensor) epicondyle of humerus. **10** Olecranon process of ulna (point of elbow). **11** Body (shaft) of ulna. **12** Lateral styloid process of ulna. **13** Body (shaft) of radius. **14** Medial styloid process of radius. **15** Radial carpal bone. **16** Ulnar carpal bone. **17** Accessory carpal bone. **18** Metacarpal bone 2. **19** Metacarpal bone 5, base. **20** Phalanges of digit 4. **21** Position of shoulder joint. **22** Elbow joint. **23** Antebrachiocarpal joint. **24** Metacarpophalangeal joint, digit 5. **25** Proximal interphalangeal joint, digit 4.

Muscles of forelimb
26 Sternocephalic muscle. **27** Sternohyoid muscle. **28** Sternothyroid muscle. **29** Cervical part of trapezius muscle. **30** Thoracic part of trapezius muscle. **31** Omotransverse muscle. **32** Latissimus dorsi muscle (important limb retractor muscle). **33-34** Ventral serrate muscle (synsarcotic muscle suspending trunk from upper end of forelimb). **33** Cervical part of ventral serrate muscle. **34** Thoracic part of ventral serrate muscle. **35** Thoracic part of rhomboid muscle. **36** Cervical part of rhomboid muscle. **37–39** Brachiocephalic muscle (main limb protractor muscle). **37** Cleidobrachial part of brachiocephalic muscle. **38** Cleidocervical part of brachiocephalic muscle. **39** Clavicular tendon in brachiocephalic muscle. **40–41** Pectoral muscles (adductor muscles of forelimb). **40** Superficial pectoral muscle. **41** Deep pectoral muscle. **42** Infraspinatus muscle. **43** Supraspinatus muscle. **44** Teres major muscle. **45** Scapular part of deltoid muscle. **46** Acromial part of deltoid muscle. **47** Biceps brachii muscle. **48** Brachial muscle. **49** Long head of triceps muscle. **50** Lateral head of triceps muscle. **51** Medial head of triceps muscle. **52** Anconeal muscle. **53** Pronator muscle of forearm. **54** Radial carpal flexor muscle. **55** Ulnar carpal flexor muscle. **56** Superficial digital flexor muscle of forelimb. **57** Superficial digital flexor tendon. **58** Deep digital flexor muscle of forelimb. **59** Deep digital flexor tendon. **60** Radial carpal extensor muscle. **61** Radial carpal extensor tendon. **62** Common digital extensor muscle. **63** Common digital extensor tendons. **64** Lateral digital extensor muscle. **65** Lateral digital extensor tendon. **66** Ulnar carpal extensor muscle (lateral ulnar muscle). **67** Oblique carpal extensor muscle and tendon. **68** Supinator muscle of forearm. **69** Abductor muscle of digit 5. **70** Interosseous muscle of digit 5. **71** Extensor retinaculum. **72** Flexor retinaculum. **73** Palmar annular ligaments.

Bones and joints of hindlimb
74 Crest of ilium. **75** Sacral tuberosity of ilium. **76** Ischiatic tuberosity (point of buttock). **77** Greater trochanter of femur. **78** Position of patella (in tendon of quadriceps femoris muscle). **79** Medial condyle of femur. **80** Lateral condyle of femur. **81** Lateral condyle of tibia. **82** Tibial crest (cranial border of tibia). **83** Body (shaft) of tibia. **84** Medial malleolus of tibia. **85** Lateral malleolus of fibula. **86** Talus. **87** Trochlea of talus. **88** Calcaneus. **89** Calcaneal tuberosity (point of hock). **90** Central tarsal bone. **91** Metatarsal bone 2. **92** Metatarsal bone 1. **93** Base of metatarsal bone 5. **94** Position of hip joint. **95** Stifle joint. **96** Crurotarsal joint. **97** Metatarsophalangeal joint, digit 2. **98** Sacrotuberous ligament (joining sacrum and 1st caudal vertebra with ischiatic tuberosity).

Muscles of hindlimb
99 Superficial gluteal muscle. **100** Middle gluteal muscle. **101** Tensor muscle of lateral femoral fascia. **102** Sartorius muscle (cranial and caudal parts). **103-105** 'Hamstring' muscles (extensor muscles of hip joint and main retractors of hindlimb). **103** Biceps femoris muscle. **104** Semitendinosus muscle. **105** Semimembranosus muscle. **106** Gemelli muscles. **107** Tendon of internal obturator muscle. **108** Quadratus femoris muscle. **109** Adductor muscles. **110-112** Quadriceps femoris muscle (major stifle extensor muscle inserting through patella and patellar tendon onto tibial tuberosity). **110** Lateral vastus muscle. **111** Medial vastus muscle. **112** Rectus femoris muscle. **113** Cranial tibial muscle and tendon. **114** Long digital extensor muscle. **115** Long digital extensor tendon. **116** Lateral digital extensor muscle and tendon. **117** Long peroneal muscle and tendon. **118** Short digital extensor muscle. **119** Lateral head of gastrocnemius muscle. **120** Deep digital flexor muscle of hindlimb (medial component). **121** Deep digital flexor muscle of hindlimb (lateral component). **122** Superficial digital flexor muscle of hindlimb. **123** Superficial digital flexor tendon. **124** Common calcaneal tendon (aggregate of structures attaching to point of hock including Achilles' tendon, superficial digital flexor tendon, accessory or tarsal tendon of 'hamstring' and gracilis muscles). **125** Accessory (tarsal) tendon (from tendinous components of biceps femoris, semitendinosus and gracilis muscles). **126** Popliteal muscle. **127** Lateral femoral fascia. **128** Crural fascia. **129** Lateral collateral stifle ligament. **130** Proximal extensor retinaculum. **131** Distal extensor retinaculum. **I-V** Digits (Digit 1 of forepaw = dewclaw).

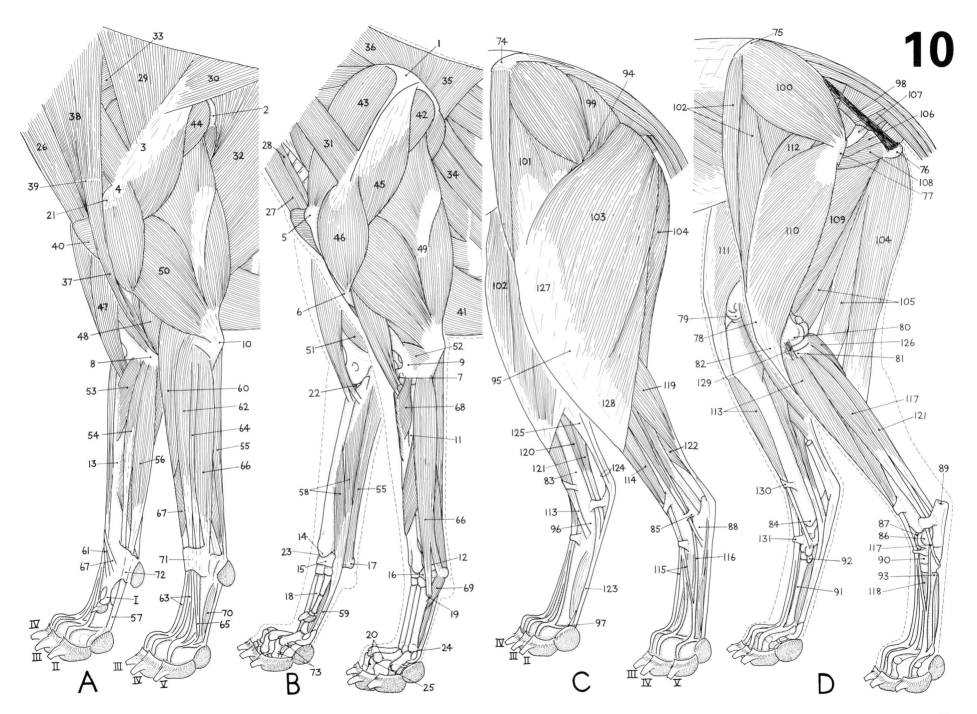

10

39

11

SURFACE FEATURES, MUSCLES AND SKELETON OF THE PAWS OF THE DOG

The selection of drawings along the top of the page are of the forepaw (**A-C** of the dorsum: **D-G** of the palmar surface). Along the bottom of the page a similar selection of hindpaw drawings are shown (**H-J** of the dorsum: **K-M** of the plantar surface). As you may see from these drawings the paws are considered together because of the many similarities between them. The final diagram at the bottom right of the page (**N**) shows a stylized view of a single digit to demonstrate the relationships between bones, tendons and ligaments. With the aid of these drawings it should be possible for you to identify many of the structural features of the paws with careful manipulation.

In the forepaw five digits (toes) are typically present although only four are fully formed — digits 2 to 5. The first digit or 'dewclaw' is always small, and in the hindpaw a dewclaw is unusual, though a feature of some breeds, the four fully formed digits being typical. Of the four main digits the central pair (digits 3 and 4) are the principal weight-bearers and so the 'axis' of the limb passes down between them.

The skeletal basis of the forepaw (**B** & **E**) consists of seven carpal bones with which five metacarpal bones articulate. In turn each metacarpal bone supports three phalangeal bones distally,

except for the dewclaw which only has two attached to a much reduced metacarpal bone. The terminal (third) phalanx in each digit is modified in shape to correspond with the capping claw. The hindpaw (**I** & **L**) also has seven tarsal bones with which the four metatarsal bones articulate. Metatarsal bone 1 if present is small and may only be attached by fibrous tissue. Digital (phalangeal) bones are identical in composition to the forepaw, although the metatarsal bones are longer than the metacarpals and somewhat narrower tending to give the hindpaw a longer and thinner shape than the forepaw.

Joints within the paws include carpals (forepaw) and tarsals (hindpaw), metacarpophalangeals (forepaw) and metatarsophalangeals (hindpaw), proximal interphalangeals and distal interphalangeals, and all function as hinges despite having anatomical structures which might suggest otherwise. The collateral ligaments associated with such hinges have already been illustrated (fig 7), and some mention has been made of the flexor and extensor muscles operating the hinged digital joints. You should recall, however, that the main movement of the paw on the limb takes place in the forepaw at the *antebrachiocarpal joint* between the radius and ulna of the forearm and the proximal row of carpal bones; and in the hindpaw at the *crurotarsal joint* between the tibia and fibula of the shank and the talus of the proximal tarsal row. Both of these joints are clearly identifiable with manipulation of the paws.

Surface views of the underside of the paws (**D** & **K**) show the position and relative size of the hairless, cushion-like **pads** — metacarpal pad in the forepaw with an equivalent metatarsal pad in the hindpaw, and digital pads in both. An additional carpal ('stopper') pad is present in the forepaw, the significance of which is difficult to suggest. The heart-shaped *metacarpal (metatarsal) pad* is the largest and lies at the junction of metacarpus (metatarsus) and the four main digits, beneath the metacarpophalangeal (metatarsophalangeal) joints which rest on it when the dog is standing up or moving around. It is the pad which consequently produces most of the ground contact surface of the paw and bears much of the dog's weight at all times. Each of the four principal digits has a smaller, oval *digital pad* beneath the distal interphalangeal joint, which support weight much as does the metacarpal (metatarsal) pad, and add to the overall area of ground contact. The *carpal pad* is peculiar to the forepaw and lies below the accessory carpal bone.

All pads have a considerably thickened, cornified epidermis, but the bulk of each one is formed from the superficial fascia beneath the dermis which has become expanded to produce an internal core of fibrous and elastic connective tissue infiltrated with large amounts of fat. Each pad is held in position by strands of anchoring connective tissue extending from the bones and tendons of the toes. Although sweat glands are distributed over the entire body, where they are essentially associated with hair follicles, opening into them separate from the sebaceous glands, the pads in the paws are really the only areas where sweat glands are at all concentrated. Here they produce a watery secretion which is unlikely to have any

role in temperature regulation. It seems possible that it may improve grip on certain surfaces but is more likely to play a part in territorial marking in some way. Some support for this latter assumption might be the way that a dog especially will scratch and paw at the ground after defaecating displacing earth for some distance.

The **claws** (nails) are composed of thickened and modified skin covering the ends of the terminal phalanges of each digit. Rather than being flaked off, as in skin generally, the outer epidermal layers in a claw remain in place and produce a solid, thickened mass. A claw has a curved, beak-like shape with a very hard wall on either side joining at a central dorsal ridge. The claw walls enclose a sole of softer tissue between their lower borders. The dorsal ridge is the thickest part and wears away less rapidly than the walls and sole, producing a permanent point to the claw. The base of a claw is tucked away beneath the ungual crest of the third phalanx and covered on its surface by a skin fold attached to the claw wall. It is only in this basal region of a claw beneath the ungual crest that the actively growing part of the claw wall is located. A claw grows down over the terminal phalanx from base to tip, much as one of your own nails grows from a nail bed. The bulk of a claw is therefore predominantly dead, horny epidermal tissue, but it is attached by a very much alive dermis to the ungual process of the phalanx. The dermis here is richly supplied with blood vessels and should the claw be torn (or overtrimmed) will bleed quite readily. Normal activity especially if on hard surfaces will tend to keep the claws worn down, although most domestic dogs will need their nails clipped at some time. The claw on the dewclaw does not touch the ground and is therefore not worn down by ground friction. Since claws are sharply curved the tip of a dewclaw might eventually come into contact with the digital pad.

A pair of **dorsal elastic ligaments** connect the ungual crest of the third phalanx with the proximal end of the second phalanx keeping the claw partially retracted by 'overextending' the distal interphalangeal joint (**N**). Active contraction in the deep digital flexor muscle will 'flex' the distal interphalangeal joint stretching the dorsal elastic ligaments and thereby 'protracting' or protruding the claws allowing them to permit better grip or to facilitate scratching.

Once the skin is removed from the paws several muscles and especially their tendons become visible. On the dorsum (**C** & **J**) extensor muscle tendons are held in place over the carpus by an *extensor retinaculum* of transversely oriented, thickened deep fascia, and over the tarsus by *proximal and distal retinacula*. On the ventrum (**F** & **M**) the tendons of the superficial and deep digital flexor muscles are prominent; ie. those muscles largely responsible for flexing (curling) the toes, and so digging the claws into the ground to obtain better grip when moving. In the forepaw the tendons of both flexor muscles begin above the carpus and run down across its rear surface into the paw held in place by a *flexor retinaculum* of deep fascia. The retinaculum converts a carpal groove into a **carpal canal** lined by deep fascia specifically for the

passage of the deep flexor tendon. The superficial wall of the canal, which would have bridged over the tendons from accessory carpal bone laterally to radial carpal bone medially, has been removed from the figure (**F**). Accompanying the tendons through the carpal canal are the main blood vessel into the paw (median artery) and the median and ulnar nerves supplying sensation to the palmar surface of the paw and pads, and innervating the intrinsic musculature.

In the hindpaw the deep flexor tendon runs across the tarsus medial to the calcaneal tuberosity in a **tarsal canal**, while the superficial tendon is incorporated in the common calcaneal tendon before it caps the point of the hock (**M**) and continues down into the metatarsus. The main blood vessel entering the hindpaw (cranial tibial artery) does not run in the tarsal canal which is on the extensor surface of the tarsus. It enters the paw by crossing the front of the tarsus where it is known as the dorsal pedal artery and reaches the underside of the paw by passing between metatarsal bones (fig 30).

The passage of tendons across front and rear of carpus and tarsus, especially where held in place by retinacula, is lubricated by **synovial tendon sheaths** (stippled on the drawings). Below the carpus and tarsus flexor tendons divide into individual elements to each of the four main digits. In addition the deep flexor tendon of the forepaw also provides a small branch to the dewclaw. Ultimately, as is indicated in the diagram at the bottom right of the page (**N**), the superficial tendons attach to the underside of the second (middle)

phalanges, while the deep tendons attach to the third (distal) phalanges.

In the second muscle drawing of the plantar surface of the forepaw (**G**), parts of both superficial and deep digital flexor tendons have been removed as far down as the lower end of the metacarpus in order to display the relationship between them at the metacarpophalangeal joints. On the fifth digit the deep flexor tendon is shown extending to its insertion on the terminal phalanx (as it is also on the third and fourth digits), but behind the metacarpophalangeal joint of digit 5 the superficial tendon is shown where it expands and completely encircles the deep tendon. Below the joint the superficial tendon continues to its insertions on the middle phalanx in a deeper position than the deep tendon (also shown diagrammatically in **N**). The two digital flexor tendons are held in place on the toes by **annular ligaments** - bands of deep fascia which attach to the phalangeal bones on either side. Digital synovial tendon sheaths around the deep and superficial flexor tendons in the toes are heavily stippled in the drawings.

Several muscles are confined to the paw itself, the intrinsic muscles, illustrated in **G**. Of these the most important seem to be **interosseous muscles** arising from the rear surfaces of the upper ends of metacarpal bones. These form most of the fleshy pad of muscle that you can palpate on the underside of the 'palm' of the paw. Each muscle runs down the back of a metacarpal bone, crosses the rear of a metacarpophalangeal joint and inserts onto the first (proximal) phalanx below the joint thus supporting it. In their passage across

the back of the metacarpophalangeal joints interosseous muscle tendons will be subjected to both friction and pressure. It was suggested earlier that sesamoid bones might develop in tendons in such situations. Hence each interosseous muscle tendon divides above the metacarpophalangeal joint and a **palmar sesamoid bone** forms in each subdivision of its tendon before they insert onto the first phalanx (shown on the second digit in **G**). Each metacarpophalangeal joint is now supported by both deep and superficial flexor tendons and by an interosseous muscle. Additional prolongations from the interosseous tendons pass dorsally onto the upper side of the digits to attach to the extensor tendon branches to each of the major digits (shown in **C**, **J** & **N**).

Whenever a paw is accepting weight during standing or moving there will be a tendency for the toes to splay apart. This is counteracted to some extent by **adductor muscles of the second and fifth digits** (**G**). These hold the outer toes adducted (pulled inwards) in relation to the axis of the limb, which you will remember passes down between the third and fourth toes.

Independent movement of the toes is strictly limited in a dog, the legs are used predominantly as supporting and propulsive elements. Should a dog use its paws to scratch itself or to dig into the ground it is still using basically the same movements as it would use when walking or running. Nevertheless, several very minor muscles are present in relation to the outer toes (digits 1 and 5) and confer a limited degree of independent movement to them. But what significance or importance these movements might have is difficult to determine. The first digit (dewclaw) has its own adductor, flexor and abductor muscles (**G**) : the fifth digit has its own abductor and flexor muscles (**G**). Of these five small muscles only the **abductor of the fifth digit** seems to have real importance in posture and movement since it is attached onto the underside of the accessory carpal bone. Onto the upper surface of the accessory carpal bone the ulnar carpal flexor muscle attaches. When the muscles acting at the carpus were considered this ulnar carpal flexor was pointed out as being one of the major supports for the carpal joint; ie. preventing its overextension and consequent collapse. Were the accessory carpal bone not firmly attached in position it could easily be displaced. A firm anchorage is assured by several ligaments running from it to the other carpal bones and to the fourth and fifth metacarpals (see fig 7), but is also aided by active contraction in the abductor of the fifth digit pulling on it in the opposite direction to the ulnar carpal flexor muscle. However, this in itself might present a problem should the accessory carpal bone be damaged (fractured possibly). Muscles pulling on the bone from above and below will mean that unless weight can be removed from the paw for some considerable time it will be difficult for a fractured accessory carpal bone to heal.

Bones and joints of forepaw
1 Antebrachium (based on radius and ulna). **2** Carpus (topographical region based on seven carpal bones). **3** Radial carpal bone. **4** Ulnar carpal bone. **5** Accessory carpal bone. **6** Carpal bone 1. **7** Carpal bone 2. **8** Carpal bone 3. **9** Carpal bone 4. **10** Metacarpus (topographical region based on metacarpal bones). **11** Metacarpal bone 1. **12** Metacarpal bone 5. **13** Carpal (stopper) pad. **14** Metacarpal pad of forepaw (metatarsal pad of hindpaw). **15** Digital pad. **16** 'Dewclaw' (1st digit of forepaw – remaining digits labelled II-V). **17** Claw (unguis). **18** Styloid process of radius. **19** Radial groove for passage of oblique carpal extensor tendon. **20** Radial groove for passage of radial carpal extensor tendon. **21** Radial groove for passage of common digital extensor tendon. **22** Styloid process of ulna. **23** Intermetacarpal spaces. **24** Proximal (1st) phalanx of digit 4. **25** Middle (2nd) phalanx of digit 4. **26** Distal (3rd or terminal) phalanx of digit 4. **27** Ungual crest of distal phalanx of digit 3. **28** Ungual process of distal phalanx of digit 3. **29-30** Sesamoid bones of metacarpophalangeal/metatarsophalangeal joints. **29** Palmar/plantar sesamoid bones of digit 3 (in tendons of insertion of interosseous muscles). **30** Dorsal sesamoid bone of digit 3. **31-33** Composite carpal joint. **31** Antebrachiocarpal joint (joint at which major carpal movement occurs). **32** Intercarpal joints. **33** Carpometacarpal joints. **34** Metacarpophalangeal joint of digit 4 of forelimb and metatarsophalangeal joint of digit 4 of hindlimb. **35** Proximal interphalangeal joint of digit 3. **36** Distal interphalangeal joint of digit 3. **37** Dorsal elastic ligaments. **38** Extensor retinaculum. **39** Digital annular ligaments (proximal, middle and distal). **40** Palmar carpal fibrocartilage (covering rear of carpus and serving as origin for most of special muscles of digits 1, 2 and 5). **41** Collateral ligaments of proximal and distal interphalangeal joints and metacarpophalangeal joint.

Muscles and tendon sheaths of forepaw
42-51 Carpal and digital extensor muscles. **42** Radial carpal extensor muscle. **43** Radial carpal extensor tendon surrounded by a synovial sheath at carpus. **44** Oblique carpal extensor muscle. **45** Oblique carpal extensor tendon surrounded by a synovial sheath at carpus. **46** Common digital extensor muscle. **47** Common digital extensor tendon surrounded by a synovial sheath at carpus. **48** Lateral digital extensor muscle. **49** Lateral digital extensor tendon surrounded by a synovial sheath at carpus. **50** Ulnar carpal extensor (lateral ulnar) muscle. **51** Ulnar carpal extensor tendon. **52** Extensions from interosseous muscles onto common extensor tendon branches. **53-60** Carpal and digital flexor muscles. **53** Superficial digital flexor muscle. **54** Superficial digital flexor tendons. **55** Deep digital flexor muscle. **56** Deep digital flexor tendon surrounded by a synovial sheath in carpal canal. **57** Digital synovial sheath around superficial and deep digital flexor tendons. **58** Ulnar carpal flexor muscle. **59** Ulnar carpal flexor tendon. **60** Radial carpal flexor tendon.

Intrinsic muscles of forepaw
61 Short flexor muscle of dewclaw. **62** Short abductor muscle of dewclaw. **63** Adductor muscle of dewclaw. **64** Adductor muscle of digit 2. **65** Adductor muscle of digit 5. **66** Flexor muscle of digit 5. **67** Abductor muscle of digit 5. **68** Interosseous muscles. **69** Tendons of interosseous muscles.

Bones and joints of hindpaw
70 Shank (crus or leg - based on tibia and fibula). **71** Tarsus (hock a topographical region based on seven tarsal bones). **72** Talus (astragalus or tibial tarsal bone). **73** Trochlea of talus. **74** Calcaneus (fibular tarsal bone). **75** Calcaneal tuberosity (point of hock). **76** Sustentaculum tali. **77** Central tarsal bone. **78** Tarsal bone 1. **79** Tarsal bone 2. **80** Tarsal bone 3. **81** Tarsal bone 4. **82** Groove on tarsal bone 4 for passage of long fibular tendon. **83** Metatarsus (topographical region based on metatarsal bones). **84** Remains of metatarsal bone 1. **85** Metatarsal bone 2. **86** Metatarsal bone 5. **87** Medial malleolus of tibia. **88** Lateral malleolus of fibula. **89** Lateral malleolar groove for passage of lateral digital extensor tendon. **90** Extensor tubercle on ungual crest of distal phalanx of digit 4. **91-93** Composite tarsal joint. **91** Crurotarsal joint (joint at which main movement at hock occurs). **92** Intertarsal joints. **93** Tarsometatarsal joints. **94** Articular capsule of tarsal joint. **95** Proximal extensor retinaculum. **96** Distal extensor retinaculum. **97** Flexor retinaculum.

Muscles and tendon sheaths of hindpaw
98-104 Tarsal flexors and digital extensors. **98** Cranial tibial muscle. **99** Cranial tibial tendon surrounded by a synovial sheath at tarsus. **100** Long digital extensor muscle. **101** Long digital extensor tendon surrounded by a synovial sheath at tarsus. **102** Lateral digital extensor tendon surrounded by a synovial sheath at tarsus. **103** Long peroneal tendon surrounded by a synovial sheath at tarsus. **104** Short digital extensor muscle. **105-108** Tarsal extensors and digital flexors. **105** Lateral component of deep digital flexor tendon surrounded by a synovial sheath at tarsus. **106** Medial component of deep digital flexor tendon surrounded by a synovial sheath in tarsal canal. **107** Superficial digital flexor tendons. **108** Common calcaneal tendon (aggregate of structures attaching to point of hock including Achilles' tendon, superficial digital flexor tendon, accessory or tarsal tendon of 'hamstring' and gracilis muscles). **109** Calcaneal bursa (between superficial flexor tendon and calcaneal tuberosity).
(C, F, J, & M after Bourdelle & Bressou, 1953; N after Evans & Christensen, 1979)

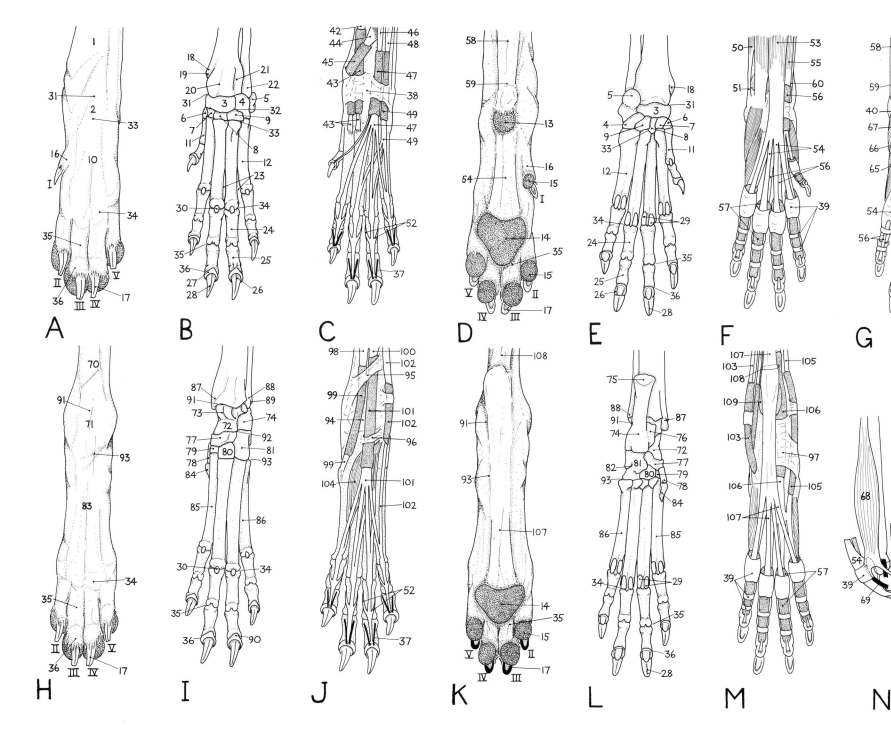

11

43

12

DEEP MUSCLES OF THE DOG – 1

These two drawings follow on from fig 9 and show further stages in the dissection and display of the axial musculature in the neck, trunk and tail – and the branchiomeric musculature in the head and throat. In **A** a number of extrinsic forelimb muscles have been removed exposing musculature of the neck and thorax while in **B** the left shoulder and arm have been removed and are placed as an inset sketch at bottom left. This removal would have been effected by cutting through the remaining extrinsic muscles attaching the forelimb to the trunk – superficial and deep pectorals, rhomboid and ventral serrate (their cut surfaces are shown by heavily stippled areas in **B**). As a result of this procedure muscles in the chest wall and at the base of the neck are now exposed, in particular the scalenes passing onto the first few ribs, and the large fan-shaped ventral serrate muscle, both of which lie in neck and thorax in the medial wall of the axilla. Removal of various muscles in the rump and thigh has exposed practically the entire extent of the abdominal wall muscles and the sacrotuberous ligament and muscles in the pelvic wall.

Bones, joints and ligaments
1 Zygomatic (supraorbital) process of frontal bone. **2** Zygomatic arch. **3** Orbit (housing and protecting eyeball). **4** Infraorbital foramen (passage of infraorbital branches of maxillary nerve and vessels). **5** Body of mandible. **6** Temporomandibular (jaw) joint. **7** External acoustic meatus (across which eardrum is stretched in life). **8** Lateral sacral crest. **9** Rib 1 (marking lateral boundary of thoracic inlet). **10** Rib 5. **11** Rib 13 (last or floating rib). **12** Dorsal (vertebral) border of scapula. **13** Caudal angle of scapula. **14** Cranial border of scapula. **15** Spine of scapula. **16** Acromion process of scapula. **17** Supraspinous fossa of scapula. **18** Scapular notch. **19** Supraglenoid tubercle of scapula. **20** Greater tubercle of humerus (point of shoulder). **21** Deltoid tuberosity of humerus. **22** Humeral condyle. **23** Olecranon process of ulna (point of elbow). **24** Shoulder joint. **25** Elbow joint. **26** Crest of ilium. **27** Coxal tuberosity of ilium (point of haunch). **28** Sacral tuberosity of ilium (point of croup). **29** Ischiatic tuberosity (point of buttock). **30** Lesser ischiatic notch. **31** Greater trochanter of femur. **32** Body (shaft) of femur. **33** Lateral condyle of femur. **34** Fabella (stifle sesamoid in tendon of origin of lateral gastrocnemius muscle). **35** Position of patella (in tendon of quadriceps femoris muscle). **36** Hip joint. **37** Stifle joint (femorotibial component). **38** Tibial tuberosity (insertion of patellar tendon). **39** Body (shaft) of tibia. **40** Medial malleolus of tibia. **41** Lateral malleolus of fibula. **42** Tarsus (ankle or hock). **43** Calcaneal bone (with calcaneal tuberosity forming point of hock).

Muscles of head
44 Orbicularis oculi muscle. **45** Levator muscle of upper lip. **46** Buccal part of buccinator muscle. **47** Molar part of buccinator muscle. **48** Superficial part of masseter muscle. **49** Deep part of masseter muscle. **50** Temporal muscle. **51** Digastric muscle. **52** Stylohyoid muscle. **53** Mylohyoid muscle. **54** Thyrohyoid muscle. **55** Cricothyroid muscle. **56** Crico-pharyngeal muscle. **57** Thyropharyngeal muscle. **58** Styloglossal muscle of tongue. **59** Sternohyoid muscle. **60** Sternothyroid muscle.

Muscles of neck, trunk and tail
61 Thoracic part of iliocostal muscle. **62** Longissimus muscle (covered by thoraco-lumbar fascia). **63** Cervical intertransverse muscles. **64** Longus capitis muscle. **65** Splenius muscle. **66** Lateral tail flexors (caudal intertransverse muscles). **67** Tail depressors (ventral sacrocaudal muscles). **68** Tail levators (dorsal sacrocaudal muscles). **69** Scalene muscle. **70** Sternocephalic muscle with sternomastoid and sternooccipital parts. **71** Cranial dorsal serrate muscle. **72** Caudal dorsal serrate muscle. **73** Cervical part of ventral serrate muscle. **74** Thoracic part of ventral serrate muscle. **75** External intercostal muscles. **76** Rectus thoracis muscle. **77** Rectus abdominis muscle. **78** Aponeurosis of origin of rectus abdominis muscle. **79** External abdominal oblique muscle. **80** Aponeurotic tendon of external abdominal oblique muscle (forming outer layer of rectus sheath). **81** Coccygeus muscle. **82** Levator ani muscle (in combination with coccygeus muscle forming pelvic diaphragm). **83** External anal sphincter muscle.

Muscles of forelimb
84 Capital part of rhomboid muscle. **85** Cervical part of rhomboid muscle. **86** Thoracic part of rhomboid muscle. **87** Omotransverse muscle. **88** Superficial pectoral muscle. **89** Deep pectoral muscle. **90** Scapular part of deltoid muscle. **91** Acromial part of deltoid muscle. **92** Supraspinatus muscle. **93** Infraspinatus muscle. **94** Teres minor muscle. **95** Teres major muscle. **96** Biceps brachii muscle. **97** Brachial muscle. **98** Long head of triceps muscle. **99** Lateral head of triceps muscle. **100** Anconeal muscle.

Muscles of hindlimb
101 Iliacus part of iliopsoas muscle. **102** Superficial gluteal muscle. **103** Middle gluteal muscle. **104** Piriform muscle. **105** Deep gluteal muscle. **106** Tendon of internal obturator muscle. **107** Gemelli muscles. **108** Quadratus femoris muscle. **109**–**112** Quadriceps femoris muscle. **109** Vastus lateralis muscle. **110** Vastus medialis muscle. **111** Rectus femoris muscle. **112** Patellar tendon (attachment of quadriceps femoris muscle onto tibial tuberosity, crossing front of stifle joint). **113**–**114** 'Hamstring' muscles. **113** Semitendinosus muscle. **114** Semimembranosus muscle. **115** Adductor muscle. **116** Cranial tibial muscle. **117** Long digital extensor muscle. **118** Long peroneal muscle. **119** Lateral digital extensor tendon. **120** Lateral head of gastrocnemius muscle. **121** Common calcaneal tendon (aggregate of tendons attaching to point of hock, including Achilles' tendon, superficial digital flexor tendon, and accessory or tarsal tendon of hamstring and gracilis muscles). **122** Deep digital flexor muscle. **123** Superficial digital flexor muscle. **124** Popliteal muscle. **125** Accessory or tarsal tendon.

Fascia and ligaments
126 Thoracolumbar fascia (deep fascia of back and loin). **127** Extensor retinacula (loops of deep fascia holding tendons in position). **128** Sacrotuberous ligament (uniting sacrum and caudal vertebra 1 with ischiatic tuberosity and forming lateral boundary of ischiorectal fossa).

Internal viscera
129 Trachea (windpipe). **130** Cranial (apical) lobe of left lung (occupying pleural pocket of left pleural cavity and projecting through thoracic inlet into base of neck). **131** Oesophagus (gullet).

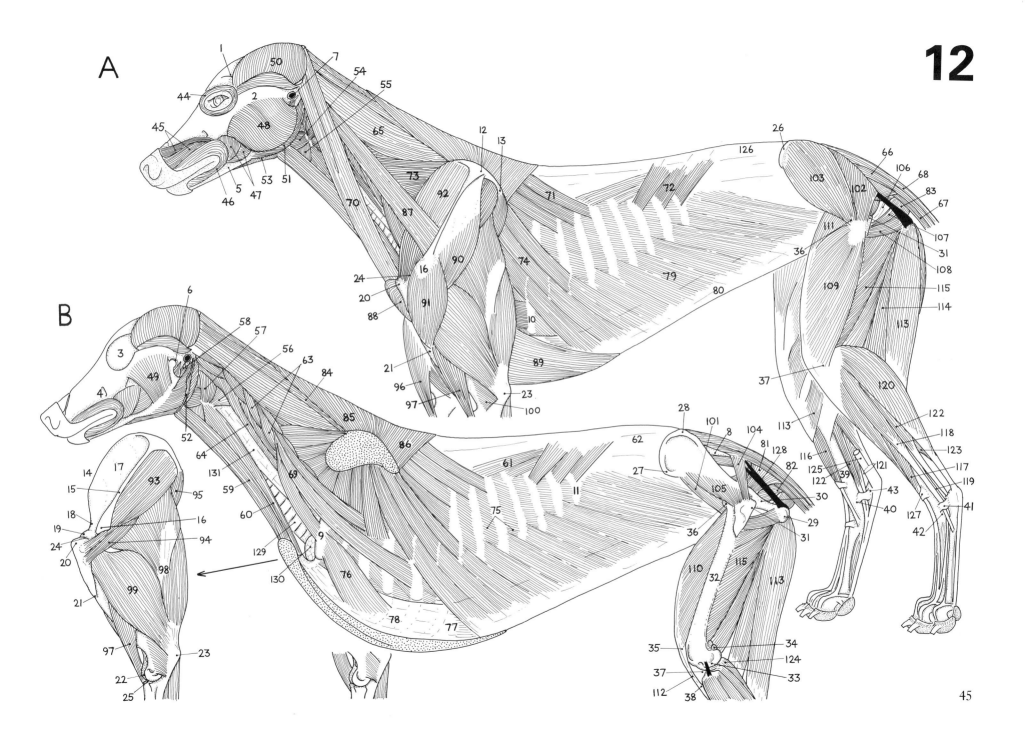

A

B

12

45

13

SURFACE FEATURES, SUPERFICIAL MUSCLES AND SKELETON OF THE DOG FROM IN FRONT

The accompanying illustrations of the dog from in front show the surface features (**A**), the superficial musculature (**D**) once the skin has been removed and the skeleton (**B**). A surface view (**C**) is also shown on which the topographical subdivisions of the body are indicated, as are the major palpable bony points that serve as 'landmarks'. Many of the surface features noticed in fig 1 are again visible in this view of a dog, particularly the jugular fossa at the base of the neck medial to the shoulder joint, and the median pectoral groove running back in the midline of the chest below the sternum.

The **jugular fossa** on either side is a triangular depression bordered by three muscles: the brachiocephalic laterally, sternocephalic medially, and the superficial pectoral muscle caudally where it extends between the sternal manubrium and the shoulder. These boundaries are palpable and you may be able to feel the external jugular vein entering the fossa *en route* for the heart in the chest. The jugular groove itself is not a particularly distinct structure especially cranially where the sternocephalic muscle crosses beneath it (**D**). It becomes more pronounced towards the jugular fossa where it is bordered above

by the brachiocephalic and below by the sternocephalic and here the vein may be distended by pressure on it immediately cranial to its entry into the jugular fossa. The internal boundary of the jugular fossa is formed from the *scalene muscle* passing from cervical transverse processes onto the lateral faces of the first few ribs. This relationship is most apparent if you take a look at fig 12A with the forelimb still in place and fig 12B after it has been removed.

The jugular fossa is normally padded out with fat and loose connective tissue. However, careful deep palpation in the fossa may reveal pulsation in the subclavian/axillary artery to the forelimb and even components of the brachial plexus of nerves against the leading edge of the first rib. It is clear from these structures that the fossa is closely related to the **thoracic inlet**, shown clearly in this view of the skeleton. The inlet is the somewhat oval bony ring formed from the first thoracic vertebra above, the short first pair of ribs and costal cartilages on either side, and the manubrium of the sternum below. As its name implies, the inlet is the entrance into the chest cavity from the base of the neck, and through it pass the structures entering or leaving the chest from the neck. The **trachea** (*windpipe*) and **oesophagus** (*gullet*) are the two largest traversing organs and they are accompanied by several blood vessels passing to and from the head and neck and also the frontlegs. Several nerves to and from the frontlegs also pass out through the inlet and then turn laterally where they are closely related to the leading edge of the first rib and its

cartilage. As we have already noticed these are the structures which might be felt deeply in the jugular fossa. Later when we deal more fully with the internal organs, we will also see that cranial extensions of the left and right pleural cavities protrude through the thoracic inlet as **pleural pockets** medial to the scalene muscles. The apex of a lung might well project into the pocket in advance of rib 1 into the base of the neck! This area of the body is clearly a complex and potentially vulnerable site.

The thoracic inlet is a fairly restricted opening with the manubrium and first thoracic vertebra being quite close together and the first ribs being short, quite straight and almost vertically oriented. Passing caudally the sternum and vertebral column diverge markedly from one another and the ribs progressively lengthen, especially beyond rib 4, with an increase in both their outward and backward curvature. Consequently as the ribcage lengthens it both deepens and widens as this view of the skeleton shows (**B**). Caudally from the manubrium the entire sternum is palpable in the midline of the chest in the median pectoral groove. This groove separates elevations formed from the pectoral muscles – large triangular masses of muscle converging on the upper end of the humerus and the crest of the greater tuberosity. Pectoral muscles are the important adductor muscles of the forelimbs keeping them in beneath the trunk to support weight most effectively. However, particularly the deep pectoral muscle also functions as a powerful limb retractor – especially pulling the limb back from a protracted position.

The **point of the shoulder** (*greater*

humeral tubercle) is in the same transverse plane as the **sternal manubrium**, lying directly lateral to it. Between these two 'landmarks', and through the brachiocephalic muscle, the *medial (lesser) humeral tubercle* is palpable where it forms the medial boundary of the *intertubercular groove*. The upper tendon of the biceps muscle is palpable in this groove and may be followed upwards across the front of the shoulder joint to its attachment on the supraglenoid tubercle of the scapula dorsomedial to the point of the shoulder. The cartilage of the humeral head is continued forwards between the two tubercles into the intertubercular groove to provide a smooth cartilage covered channel in which the biceps tendon runs. The shoulder joint capsule is also extended into the groove and, although generally referred to as an intertuberal or **bicipital bursa** in relation to the tendon, it does in fact enclose the tendon in a tubular synovial sheath cushioning and lubricating its passage. The tendon and encircling sheath are held in place in the groove by a transverse humeral retinaculum of ligamentous tissue joining the tubercles. The bulk of the biceps muscle can be palpated in the arm although covered by pectoral muscles for much of its length. Its lower tendon of insertion can be traced into the forearm close to the elbow joint where it attaches in conjunction with the brachial muscle.

Medial to the scapula, the shoulder joint and the upper arm, is the **axilla** roughly equivalent to your own 'armpit'. Although potentially a large space in actual fact its volume is very small. It is a very narrow fascia lined space between the lateral thoracic wall (mainly the ventral serrate, scalene and rectus thoracic muscles as you may see in fig 12B), and the medial side of the shoulder and arm (subscapular, teres major and triceps muscles). Ventrally it is closed by the deep pectoral muscle running between the undersurface of the chest and the inner surface of the arm. Its volume is small since the medial and lateral walls are pressed against each other when the limb is not abducted, compressing a modicum of loose spongy connective tissue and fat and obliterating the space. Cranially the axilla is in continuity with the jugular fossa beneath the brachiocephalic muscle: caudally it extends beneath the latissimus dorsi muscle and the deep pectoral muscle, and where these diverge is closed in by the cutaneous muscle of the trunk. The 'armpit' is not as extensive as in yourself since the dog cannot abduct its limb away from its chest to the same extent as you.

Apart from the loose spongy tissue the principal contents of the axilla are the blood vessels and nerves of the forelimb (the *axillary artery and vein*, and the nerves arising from the *brachial plexus*). These are passing out through the thoracic inlet ventral to the scalene muscle around the leading edge of rib 1 and across the axilla into the shoulder and arm. Further caudally in the axilla related to ribs 2 or 3, the *axillary lymph node* is located. This node is occasionally accompanied by an accessory axillary node although it is not normally possible to feel either node.

An examination of the jugular fossa and axillary region brings to mind again that the shoulder girdle (scapula) only has a muscular connection with the trunk (a **synsarcosis**). In ourselves a collar bone (clavicle) crosses the equivalent of the jugular fossa, and the depression you can feel in yourself above your collar bone is the supraclavicular fossa. The principal muscle of attachment in the synsarcosis is the large, fan-shaped **ventral serrate** passing upwards from the lower ends of the first eight ribs and the transverse processes of the last five neck vertebrae, to converge on the upper end of the internal face of the scapula. It suspends the trunk from the upper end of the limb and in combination with its fellow of the opposite side it forms a sling for the thorax. Several subsidiary muscles assist the ventral serrates in holding the scapula in place against the chest wall: dorsally the *trapezius and rhomboid* muscles; cranially the *omotransverse* and *brachiocephalic muscles*; ventrally the *superficial* and *deep pectoral muscles*; and caudally the *latissimus dorsi*. This collection of muscles constitutes the extrinsic musculature of the limb which are not only important in a postural capacity but are the major muscles moving the limb on the trunk.

Because of this flexible muscular union, and through the action of the muscles making it up, the scapula is capable of a number of movements relative to the chest wall. It can 'rotate', pivoting about a transverse horizontal axis passing through the origin of the ventral serrate muscle on the ribs. Such an activity moves the ventral (articular) angle of the scapula backwards and forwards and so aids in limb movement and especially in lengthening the stride.

As a dog lengthens its stride when it is running at increasing speeds, its scapula also moves backwards and forwards on the chest wall. These slight gliding movements occur in conjunction with the rotatory movements just alluded to and are due to activity in the same muscles.

The scapula may also be abducted and adducted in relation to the chest wall. Such movements, although limited, are very necessary in balance maintenance especially when an animal is changing direction. Abduction involves pulling the upper end of the scapula in towards the midline of the withers (trapezius and rhomboid muscle activity) promoting outward movement of the lower (articular) end, effectively opening out the axilla. However, abduction will occur simply as a result of body weight tending to splay the legs apart. This tendency must be continually counteracted primarily by activity within the pectoral muscles. Scapular adduction (inward movement of the articular angle) is therefore a continuous function of the pectoral muscles whenever the dog is on its feet.

Although covered by the *sternohyoid and sternothyroid muscles* on the underside of the neck, the **trachea** is clearly palpable for practically its entire length in the neck because of its cartilaginous composition. The oesophagus, however, without a firm framework is not normally discernible. We will see later (fig 32) that lymph nodes are located in certain strategic positions in the body. The *superficial cervical lymph nodes* are quite a prominent pair cranial to the scapula on the side of the neck. Although lying under the brachiocephalic and omotransverse muscles these nodes may often be felt, especially so if enlarged for any reason.

Lower down the limb on the cranial (flexor) surface of the elbow you will be able to discern/palpate a roughly triangular depression, the **cubital fossa**. Its base is on a line joining the humeral epicondyles proximal to the elbow joint, both of these you may readily palpate. The medial boundary is the *pronator teres muscle* passing into the forearm onto the radius from the medial epicondyle (see fig 10A): the lateral boundary is the *radial carpal extensor muscle* from the lateral epicondyle forming the cranial contour of the forearm. Entering the fossa medially are the closely related tendons of insertion of the biceps and brachialis muscles which are clearly palpable attaching to the radius and ulna. The floor of the fossa is formed laterally from the supinator muscle of the forearm (see fig 10B) and medially by the fibrous layer of the elbow joint capsule. The fossa is roofed over by the deep fascia of the forearm which you will remember we specified as being a prominent fascial sheet. Like the jugular fossa the cubital contains fat and loose connective tissue and has several vessels and nerves passing through it. Running in close association are the *brachial artery and vein* and the *median nerve* all disappearing into the forearm beneath the pronator muscle. Pulsation in the brachial artery may be felt on compression against the humeral condyle. A median cubital vein enters the fossa medially joining the cephalic vein in the roof with the brachial vein in the floor.

The cubital fossa has its equivalent depression in the hindlimb – the **popliteal fossa** – a depression caudal to the stifle joint; ie. on its flexor surface (see fig 16). The divergent hamstrings, semitendinosus medially and biceps femoris laterally, and the gastrocnemius muscle below form the boundaries of the fossa. The fossa is deep and its floor is formed from the popliteal surface of the femur, the stifle joint capsule, and the deep fascial covering on the popliteus muscle. Embedded in fat within the fossa is the prominent *popliteal lymph node*, a structure which is normally palpable. The lateral saphenous vein enters the fossa from the lower leg, and the popliteal artery and vein and the tibial and common fibular nerve traverse the fossa more deeply.

Surface features and topographical regions
1 Nasal plane (pigmented hairless skin). **2** External nostril (leading into nasal vestibule, movable part of nose surrounded by nasal cartilages). **3** Philtrum. **4** Muzzle. **5** Lips. **6** Commissure of lips at angle of mouth. **7** Chin (mentum). **8** Tongue. **9** Cheek (based on buccinator muscle). **10** Foreface. **11** Stop. **12** Forehead. **13** Pinna of external ear (based on auricular cartilage). **14** Throat. **15** Jugular groove (containing external jugular vein). **16** Jugular fossa (triangular depression at base of neck communicating internally with axilla). **17** Dorsal neck region. **18** Lateral neck region. **19** Tracheal region. **20** Median pectoral groove. **21** Presternal region (breast based on superficial pectoral muscles. **22** Sternal region (brisket based on deep pectoral muscles. **23** Axillary region (includes axilla or axillary fossa – the armpit). **24** Scapular region (shoulder). **25** Shoulder joint region. **26** Brachial region (brachium, arm or upper arm). **27** Cubital region (including cubital fossa). **28** Antebrachial region (forearm). **29** Carpal region (carpus or wrist based on carpal bones and joints). **30** Metacarpal region (front pastern). **31** Phalangeal region (digits or toes). **32** Claws (capping ungual processes of distal phalanges). **33** Digital pad (beneath distal interphalangeal joint). **34** 'Dewclaw' (1st digit of forepaw – remaining digits are designated by roman numerals II–V). **35** Interdigital space.

Bones of skull
36 Bony nasal opening (piriform aperture leading into bony part of nasal cavity bounded by nasal processes of incisive bones). **37** Nasal bone. **38** Infraorbital foramen (passage of infraorbital branches of maxillary artery and nerve). **39–42** Mandible (lower jaw). **39** Body of mandible. **40** Angular process of mandible. **41** Ramus of mandible. **42** Mental foramen (passage of mental branches of mandibular alveolar nerve and vessels). **43** Mandibular symphysis (fibrocartilaginous intermandibular joint allowing little if any movement). **44** Jaw (temporomandibular) joint. **45** Zygomatic arch (bridge of bone connecting face and cranium below eye). **46** Orbit (housing and protecting eyeball). **47** Zygomatic (supraorbital) process of frontal bone.

Vertebral column, ribs and sternum
48 Cervical vertebra 7. **49** Transverse process of cervical vertebra 6. **50** Rib 1 (short and fairly straight). **51** Costal cartilage of rib 1 (articulating with manubrium of sternum). **52** Manubrium of sternum (sternebra 1 elongated into base of neck). **53** 2nd sternebra (sternum formed from 8 sternebrae joined by intersternebral cartilages which tend to ossify in aged dogs). **54** Thoracic inlet (more or less oval opening bounded by sternal manubrium, first pair of ribs and thoracic vertebra 1).

Forelimb skeleton and joints
55 Dorsal (vertebral) border of scapula. **56** Cranial border of scapula. **57** Spine of scapula. **58** Acromion process of scapula. **59** Supraglenoid tuberosity of scapula (attachment for biceps brachii muscle). **60** Greater tubercle of humerus (point of shoulder). **61** Deltoid tuberosity of humerus. **62** Crest of greater tubercle of humerus. **63** Intertubercular (bicipital) groove of humerus (for passage of biceps tendon surrounded by bicipital bursa and held in place by a transverse humeral retinaculum). **64** Lesser tubercle of humerus. **65** Body (shaft) of humerus. **66** Supratrochlear foramen of humerus. **67** Capitulum of humeral condyle (small lateral part of condyle articulating with radial head). **68** Trochlea of humeral condyle (pulley-shaped medial part of condyle articulating with trochlear notch of ulna). **69** Medial (flexor) epicondyle of humerus. **70** Lateral (extensor) epicondyle of humerus. **71** Olecranon process of ulna (point of elbow). **72** Body (shaft) of ulna. **73** Lateral styloid process of ulna. **74** Head of radius. **75** Neck of radius. **76** Lateral tuberosity of radius. **77** Body (shaft) of radius. **78** Medial styloid process of radius. **79** Grooves on surface of distal end of radius for passage of extensor tendons (medial groove for oblique carpal extensor: middle groove for radial carpal extensor: lateral groove for common digital extensor). **80** Carpal bones (radial carpal and ulnar carpal in proximal row, carpal bones 1–4 in distal row). **81** Metacarpal bones (5 with metacarpal bone 1 reduced in size). **82** Lateral surface of base of metatarsal bone 5. **83** Phalanges of digits. **84** Shoulder joint. **85** Elbow joint (composite synovial joint – 3 parts contained in a single joint capsule). **86** Antebrachiocarpal joint (main component of carpal joint and joint at which most movement occurs). **87** Carpometacarpal joints. **88** Proximal interphalangeal joints.

Muscles
89 Levator muscle of nostril wing and upper lip. **90** Levator muscle of upper lip. **91** Orbicularis oris muscle of lips (sphincter muscle of mouth opening). **92** Platysma muscle. **93** Zygomatic muscle (retractor muscle of angle of mouth). **94** Orbicularis oculi muscle of eyelids (sphincter muscle of palpebral fissure). **95** Frontal muscle. **96** Rostral auricular muscles. **97** Sternohyoid muscle. **98** Sternocephalic muscle. **99** Omotransverse muscle. **100** Cervical part of trapezius muscle. **101** Superficial pectoral muscles. **102** Deep pectoral muscle. **103** Cleidocervical part of brachiocephalic muscle. **104** Cleidobrachial part of brachiocephalic muscle. **105** Clavicular tendon within brachiocephalic muscle (representing fibrous remnants of clavicle). **106** Acromial part of deltoid muscle. **107** Lateral head of triceps muscle. **108** Brachial muscle (elbow joint extensor occupying brachial [musculospiral] groove of humerus). **109** Biceps brachii muscle (shoulder joint extensor and elbow joint flexor inserting onto both radius and ulna in forearm). **110** Radial carpal extensor muscle and tendon. **111** Common digital extensor muscle and tendon. **112** Lateral digital extensor muscle. **113** Oblique carpal extensor muscle. **114** Pronator muscle of forearm.

Blood vessels
115 Linguofacial vein. **116** Maxillary vein. **117** External jugular vein (entering jugular fossa at base of neck *en route* for thoracic inlet). **118** Dorsal common digital veins. **119** Cephalic vein. **120** Axillobrachial vein (connection of cephalic vein with brachial/axillary vein in axilla). **121** Omobrachial vein (minor superficial connection between cephalic vein and external jugular vein).
(After Ellenberger, Dittrich & Baum, 1956)

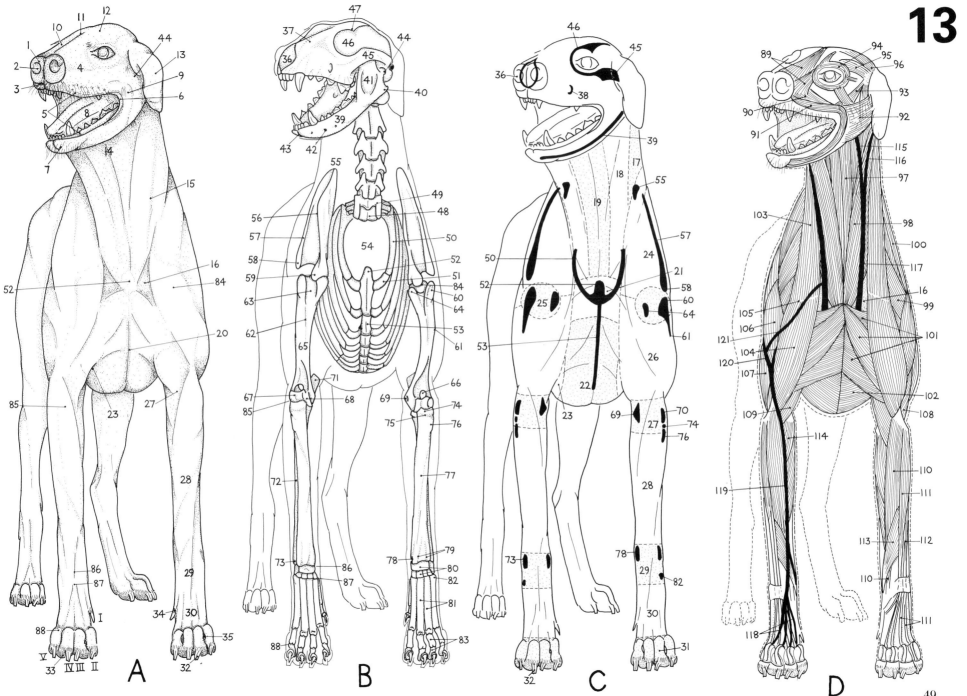

13

49

14

DEEP MUSCLES OF THE DOG AND BITCH – 2

In this second pair of deep muscle drawings exposure of the musculature of the head, neck, trunk and tail is continued from fig 12B. In the head the remaining facial musculature has been removed exposing the branchiomeric (modified visceral) muscle of the jaws, tongue and pharynx. Removal of the ventral and dorsal serrate muscles and thoracolumbar fascia, all in position in **A**, exposes the three main components of the epaxial musculature – iliocostal and longissimus components in the back and loins, and the spinal and semispinal parts of the transversospinal component in the back and neck. Additional removal of the splenius from the neck exposes the cervical longissimus and the biventer and complexus parts of the capital semispinal muscle.

The external abdominal oblique muscle has been removed from the caudal end of the chest and flank exposing the rectus abdominis muscle as it runs forwards on the underside of the belly onto the chest, and the internal abdominal oblique lying in the flank and belly. Removal of rump and thigh musculature has also exposed the iliopsoas muscle emerging from the abdomen behind the caudalmost extent of the internal abdominal oblique muscle.

In **B** the internal abdominal oblique muscle has been removed exposing the transverse abdominal in the flank and practically the entire length of the rectus abdominis in the belly. In the chest removal of the scalene, rectus abdominis and external intercostal muscles exposes the ribs, costal cartilages and internal intercostal muscles. Consequently in the trunk overall, only the innermost layers of the body wall muscles (internal intercostals in the thorax between the ribs, transverse abdominal in the abdomen) are left in place on the surface of the serous membranes lining the body cavities.

In the pelvic wall complete removal of the rump and thigh muscles and the femur exposes the hip bone and sacrotuberous ligament. Medial to this ligament the 'pelvic diaphragm' is displayed composed of the coccygeus and levator ani muscles. This diaphragm originating from the hip bone and inserting onto caudal vertebrae and blending caudally with the external anal sphincter, separates off an ischiorectal fossa laterally between it and the sacrotuberous ligament.

Bones, joints and ligaments
1 Body of mandible. **2** Angular process of mandible. **3** Coronoid process of mandible (removed in B). **4** Temporomandibular (jaw) joint. **5** External occipital protuberance (occiput – most dorsocaudal portion of cranium). **6** Nuchal crest (division between dorsal and caudal surface of cranium). **7** Zygomatic arch (sawn through and removed). **8** Temporal fossa (origin of temporal muscle). **9** External acoustic meatus (across which eardrum stretched in life). **10** Basihyoid bone. **11** Thyroid cartilage of larynx. **12** Lateral sacral crest (fused 2nd and 3rd sacral transverse processes). **13** Rib 1. **14** Rib 13 (last or floating rib). **15** Costochondral junction between bony ribs and costal cartilages. **16** Costal arch (fused costal cartilages of ribs 10–12 [asternal ribs], linked by fibrous tissue with costal cartilage of rib 9, last sternal rib). **17** Manubrium of sternum (sternebra 1 enlarged into base of neck). **18** Wing of ilium. **19** Crest of ilium. **20** Coxal tuberosity of ilium (point of haunch). **21** Sacral tuberosity of ilium (point of croup). **22** Greater ischiatic notch of hip bone. **23** Lesser ischiatic notch of hip bone. **24** Ischiatic tuberosity (point of buttock). **25** Ischiatic spine. **26** Pubic pecten of hip bone. **27** Obturator foramen. **28** Acetabulum of hip joint. **29** Hip (coxofemoral) joint. **30** Greater trochanter of femur. **31** Sacrotuberous ligament (joining lateral sacral crest and 1st caudal transverse process with ischiatic tuberosity).

Muscles of head
32 Temporal muscle. **33** Medial pterygoid muscle. **34** Geniohyoid muscle. **35** Mylohyoid muscle. **36** Thyrohyoid muscle. **37** Jugulohyoid muscle. **38** Cricothyroid muscle. **39** Cricopharyngeal muscle. **40** Thyropharyngeal muscle. **41** Hyopharyngeal muscle. **42** Ceratopharyngeal muscle. **43** Styloglossal muscle. **44** Hyoglossal muscle.

Muscles of neck, trunk and tail
45 Sternothyroid muscle. **46** Lumbar part of iliocostal muscle. **47** Thoracic part of iliocostal muscle. **48** Lumbar part of longissimus muscle. **49** Thoracic part of longissimus muscle. **50** Cervical part of longissimus muscle. **51** Capital part of longissimus muscle. **52** Thoracic part of spinal muscle. **53** Cervical part of spinal muscle. **54** Thoracic part of semispinal muscle. **55** Capital part of semispinal muscle (biventer). **56** Capital part of semispinal muscle (complexus). **57** Cervical intertransverse muscles. **58** Longus capitis muscle. **59** Longus colli muscle. **60** Splenius muscle. **61** Scalene muscle. **62** Tail depressors (ventral sacrocaudal muscles). **63** Tail levators (dorsal sacrocaudal muscles). **64** Lateral tail flexors (caudal intertransverse muscles). **65** Ventral serrate muscle (remains of). **66** Cranial dorsal serrate muscle. **67** Caudal dorsal serrate muscle. **68** External intercostal muscles. **69** Internal intercostal muscles. **70** Internal abdominal oblique muscle. **71** Tendon (aponeurosis) of internal abdominal oblique muscle (contributing to external layer of rectus sheath). **72** Inguinal ligament. **73** Transverse abdominal muscle. **74** Tendon of transverse abdominal muscle (contributing to external layer of rectus sheath). **75** Rectus abdominis muscle. **76** Tendinous intersections in rectus abdominis muscle. **77** Prepubic tendon. **78** Coccygeus muscle. **79** Levator ani muscle. **80** External anal sphincter muscle. **81** Ischiocavernosus muscle. **82** Bulbospongiosus muscle. **83** Constrictor muscle of vestibule. **84** Constrictor muscle of vulva. **85** Thoracolumbar fascia.

Limb muscles
86 Symphyseal tendon (midline fibrous plate attached to pelvic symphysis providing origin for medial thigh muscles). **87** Iliacus part of iliopsoas muscle. **88** Psoas major part of iliopsoas muscle. **89** Superficial pectoral muscle. **90** Deep pectoral muscle.

Viscera
91 Trachea (windpipe). **92** Cranial (apical) lobe of left lung. **93** Oesophagus (gullet). **94** Rectum (continuation of descending colon within pelvis). **95** Vaginal process (extension of parietal abdominal peritoneum through inguinal canal containing testis). **96** Scrotum (sparsely haired skin sac containing vaginal processes and testes). **97** Penis. **98** Prepuce (sheath). **99** Vestibule. **100** Vulva. **101** Position of teats of mammary glands.

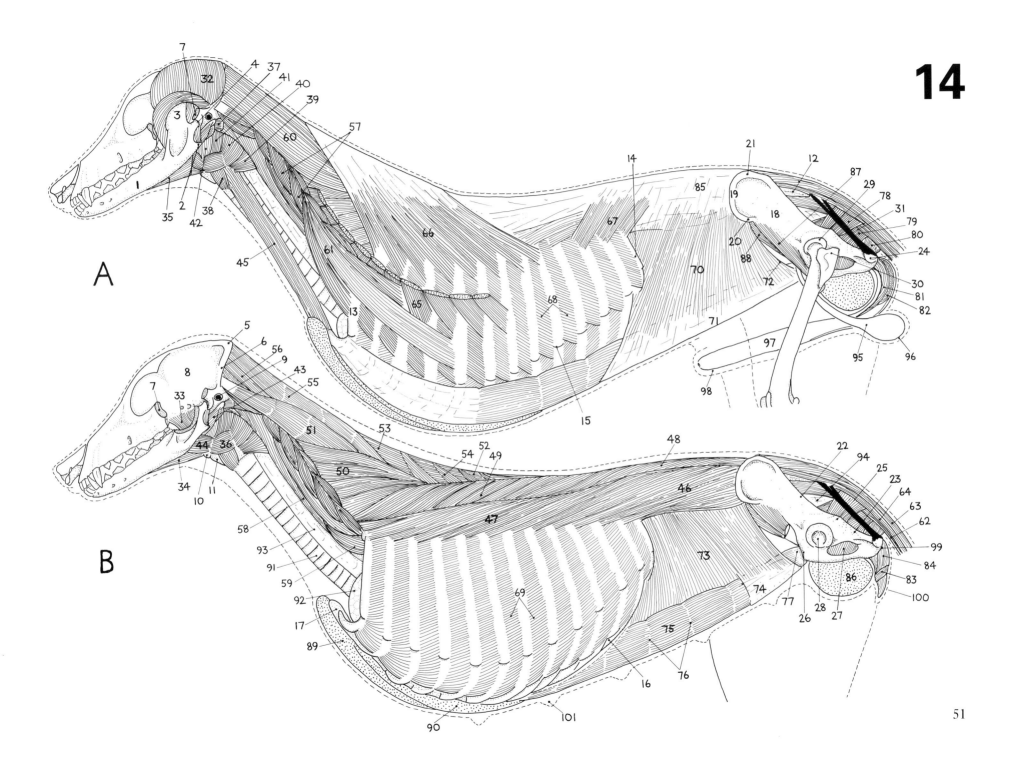

14

A

B

15

SURFACE FEATURES, TOPOGRAPHICAL REGIONS AND MUSCULOSKELETAL SYSTEM OF THE DOG FROM ABOVE

These five illustrations of the dog from above show surface and skeletal views, and various layers of dorsal musculature. The upper body contour, based upon the neck, back, loin and rump, is shown in **A** from the surface. The skeleton is drawn in **B** and a second surface view (**C**) is given on which the major palpable features are inked in and the topographical regions indicated. As you can see from this third drawing the summits of the spinous processes of vertebrae are palpable in the dorsal midline. However, these are mainly apparent in the back, loin and rump, since they lie some way below the surface for most of the neck. Capping spinous processes and joining them together in the midline, the **supraspinous ligament** runs the length of the trunk. From the summit of the first thoracic spinous process the ligament is continued as the **nuchal ligament** which attaches to the spine of the axis palpable at the cranial end of the neck. The nuchal ligament itself cannot be palpated as a distinct structure. Ribs can be felt in the chest, although not close to the dorsal midline where they are buried deeply beneath epaxial muscles. The tips of some transverse processes of cervical and loin vertebrae are also palpable. At the front end of the trunk the shoulder blades (scapulae) can be identified, their vertebral borders lying on either side of the withers.

In the pelvic region at the rear end of the trunk the crests of the iliac bones can be felt on either side of the midline. The upper end of a crest is the sacral tuberosity (point of the croup), the lower end, the coxal tuberosity (point of the haunch), but neither is particularly prominent. What we are calling a tuberosity here is strictly only the cranial component of a more extensive area. The tuberosities include cranial and caudal spines and the intervening border of the ilium. Thus the sacral tuberosity is composed of a cranial dorsal iliac spine a caudal dorsal iliac spine and the intervening dorsal iliac border. As you will see a transverse line joining the sacral tuberosities of left and right sides crosses the backbone at about the lumbosacral junction.

It is unlikely that you will be able to identify any further bony features of the spine from the surface since for its entire length the vertebral column is buried beneath a stout mass of dorsal (epaxial) muscles most prominent in the loins. The epaxial musculature is closely invested by a dense sheet of deep **thoracolumbar fascia**. Once the skin and superficial fascia have been removed this can be seen in **D** and **E** to form a very close investment to the muscles of the back and loin and is continued over the rump as *gluteal fascia*. In addition it acts as a tendon of attachment for several body muscles notably the latissimus dorsi and thoracic trapezius which we have considered with the forelimb muscles, and also for the oblique abdominal muscles which we will be considering shortly. The fascia is securely attached to the spinous processes of thoracic and abdominal vertebrae, actually blending with the supraspinous ligament uniting their summits. It also has attachments to the tips of lumbar transverse processes and ribs lateral to the dorsal muscles. Sheets of fascia (intermuscular septa) extend inwards from the thoracolumbar fascia passing between dorsal muscles to attach to the spines, arches and transverse processes of all of the vertebrae.

Without the aid of muscles the spine made up from a string of individual vertebral bones (joined together by intervertebral discs, synovial joints and longitudinal ligaments) does not have any natural rigidity and is incapable of independent movement. Spinal stability and mobility are therefore both produced by axial muscle. **Epaxial (dorsal) muscles** consist of many overlapping parts lying above the level of transverse processes. They extend the entire length of neck, trunk and tail and appear as a more or less continuous block filling the trough that exists between transverse and dorsal spinous processes. The epaxial muscles are gradually displayed in the drawings on the right of the page (**E, F & G**) but they are also progressively displayed from the side later on in figs 20, 21 & 22. Three broad groups of epaxial musculature are present. The most lateral of the epaxial muscles is the **iliocostal muscle** in the loin and thorax. This is important topographically since it forms the ventral boundary of the epaxial muscles in the chest. The middle component, the **longissimus muscle**, is the largest, extending the entire length of the trunk and neck up to the rear of the skull and even having continuations caudally into the tail as *dorsal sacrocaudal muscles*. The medial component, the **transversospinal muscle**, is the most complex and consists of a number of muscles including the *multifidus, rotators, interspinals, semispinal and spinal muscles*. As you can see from all of these drawings, epaxial muscle is exceedingly complex in its detailed arrangement since it is made up of numerous small components spanning only a few vertebrae at a time.

Acting across intervertebral joints, epaxial muscles contract in cooperation with hypaxial muscles to stabilize the column in the trunk when a dog is standing up so that weight can be suspended from it. When a dog is moving they can act as spinal extensors, straightening the back and loins and raising the head and neck. If acting on one side only they will bend the backbone to that side. The **hypaxial muscles** lie below the level of transverse processes and extend down and around the body cavities of the thorax and abdomen to the ventral midline, easily recognizable as the musculature of the flanks and belly walls. However, in the thorax the lateral musculature is somewhat more complex because of the presence of ribs, and in the pelvis it is partially interrupted but still represented as the muscles of the pelvic diaphragm.

Hypaxial musculature is also present as *subvertebral musculature* immediately beneath the column, of special import in the neck and loins. Epaxial musculature above the vertebral column cooperates with the antagonistic subvertebral (hypaxial) musculature below it. But in many activities abdominal muscles or even the force of gravity will provide an antagonist to epaxial muscles, consequently the subvertebral muscles are smaller than the epaxials. Nevertheless subvertebral muscles do attain some considerable size in the loins, especially the *quadratus lumborum* extending from the last few thoracic vertebral bodies to the transverse processes of lumbar vertebrae and the iliac wing, and are important in either flexing or fixing the vertebral column in the loins.

Abdominal wall muscles are flattened sheets in the flanks and belly superimposed on each other. The muscles of left and right sides are attached above to the thoracolumbar fascia, lumbar vertebrae and ribs and meet below in the **linea alba** (white line) of the ventral midline of the belly. This fibrous union extends back from the xiphoid cartilage of the sternum to the prepubic tendon and cranial pubic ligament of the pelvic girdle. The two outer layers of muscle are termed oblique which is a reference to the orientation of their muscle fibres: in the external layer they pass caudoventrally, in the internal layer cranioventrally. The innermost muscle layer of the abdominal wall has transversely arranged fibres while the rectus abdominis muscles, one on either side of the white line, extending from attachments on sternum and ribs to the prepubic tendon on the cranial border of the pubic bones, have longitudinally oriented fibres. The fibres in the two oblique sheets and those in the transverse/rectus component are arranged at right angles to each other, an arrangement which presumably adds strength

to the wall.

The abdominal wall muscles are particularly important muscles and participate in a variety of activities which will include such things as flexing the backbone (arching the back) when the muscles contract simultaneously on both sides, and lateral flexion of the backbone when the muscles of one side only contract to curl the body up. In both of these actions the abdominal muscles supplement the dorsal epaxial and subvertebral muscles. Since the muscles of either side meet below in the midline of the belly they form a sling suspended from the lumbar vertebral column. This sling provides support for abdominal organs primarily the guts but also the uterus, particularly of a pregnant bitch, and the distended bladder. It will also be able to compress these abdominal contents, raising pressure inside the abdomen when it is necessary to assist in straining in important functions such as defaecation, urination and parturition. Expiration, the act of breathing out, is also assisted by the press, displacing abdominal organs forwards in the abdominal cavity, pushing them up against the diaphragm.

Aside from many of the more superficial muscles of the thoracic wall such as the pectorals, latissimus dorsi and ventral serrates, which we have already mentioned in association with their actions around the shoulder, several other chest muscles are more concerned with the ribcage in breathing movements: *inspiratory*, causing air to be drawn into the lungs by increasing the size of the thorax; *expiratory*, forcing air out of the lungs by decreasing the thoracic volume.

Ribs are attached at their upper ends to the vertebral column (costovertebral joints) and at their lower ends directly or indirectly to the sternum (costosternal joints). Ligaments placed above and below these joints indicate that they are confined to a single plane of movement; ie. moving fore-and-aft about a vertical axis through their extremities. An analogy has often been drawn between this type of rib movement and the movement of a bucket handle. On breathing in ribs rotate outwards and forwards, equivalent to lifting a bucket handle away from the rim; on breathing out the ribs rotate backwards and inwards.

Inspiratory rib movements are in large measure due to activity in the **external intercostal muscles** filling the spaces between ribs, their fibres running from rib to rib in a caudoventral direction (downwards and backwards). When they contract they tend to pull the more caudally placed rib forwards and outwards in relation to the more cranial one. For inspiratory rib movement to occur with any level of efficiency the first few ribs must be prevented from being pulled backwards to provide a stable fulcrum about which overall rib movement can occur. This is presumably accomplished by *scalene muscles* extending back from cervical vertebrae onto the first few ribs. Rib movements therefore increase the volume of the thoracic cavity essentially by increasing its width. Although rib movements are important to a dog in breathing, particularly when it is running around and exercising, the major breathing muscle at all times is the **diaphragm**, the sheet of muscle which separates thoracic and

abdominal cavities filling the thoracic outlet.

Breathing out is essentially a passive act relying on the natural elasticity of lung tissue − stretched lungs tend to recoil when inspiratory muscles relax. During more rapid breathing a dog must make definite expiratory efforts since simple elastic rebound of stretched lungs cannot clear them of stale air rapidly enough. Expiratory effort is produced by abdominal wall musculature squeezing the abdomen and pushing abdominal organs forwards against the diaphragm decreasing the thoracic volume. At the same time, since abdominal muscles are attached to the ribcage, they will automatically pull the ribs caudally.

The wall of the pelvis is formed predominantly from hip bones with some contribution from the sacrotuberous ligament and from rump muscles originating from the ligament. Nevertheless the pelvic wall is incomplete laterally and dorsolaterally and so helping to confine pelvic viscera in this area is a muscular '**pelvic diaphragm**' consisting of the *coccygeus and levator ani muscles*. Strictly these muscles lie *in* the pelvic cavity and form a partition between the terminal ends of the digestive and urogenital tracts medially and an ischiorectal fossa laterally. The main importance of this diaphragm lies in its contribution to maintaining the pelvic organs in position and its assistance in defaecation. Both muscles are sheet-like components extending from the hip bone (ischiatic spine for coccygeus, inner surface of pubis and entire pelvic symphysis for levator ani) to the transverse processes and bodies of tail vertebrae back to six or seven. The levator ani also has a significant attachment to/association with the external anal sphincter muscle around the anal canal effectively sealing off the pelvic cavity caudally beneath the perineum. It seems fairly obvious that these muscles will cooperate with tail depressors to pull the tail down against the anus and external genitalia, and between the hindlegs. However, during defaecation these pelvic diaphragm muscles are important in squeezing the rectum and anal canal to void the faeces. At such a time the tail is raised to avoid becoming soiled, hence these diaphragm muscles, acting at the same time as tail levators, produce the pronounced kink in the tail at about the sixth or seventh vertebra which is often apparent in the defaecating dog.

Also of importance during the straining accompanying voiding is the role played by these same pelvic diaphragm muscles in stopping pelvic organs from being pushed backwards. Straining results from raising pressure within the abdomen which in turn raises pressure inside the pelvis (the two cavities being continuous). Increased intrapelvic pressure compresses the rectum and forces faeces through the anal canal. The coccygeus and levator ani muscles help to prevent the intestine from herniating through the pelvic wall lateral to the anus into the ischiorectal fossa.

Surface features and topographical regions
1 Upper eyelid (supporting cilia). **2** Foreface (dorsal nasal region). **3** Muzzle (lateral nasal region). **4** Stop. **5** Forehead. **6** Temporal region. **7** Frontal (supraorbital) region. **8** Parietal region. **9** Occipital region (occiput).

10 Auricular region (pinna of external ear). **11** Neck. **12** Crest of neck. **13** Dorsal neck region. **14** Interscapular region (withers). **15** Thoracic vertebral region (back). **16** Lumbar region (loins). **17** Sacral region (croup). **18** Clunial region (includes ischiorectal fossa). **19** Root of tail. **20** Caudal region (tail). **21** Scapular region (shoulder). **22** Costal region (chest or thorax). **23** Lateral abdominal region (flank). **24** Coxal tuberosity region (haunch). **25** Gluteal region (rump). **26** Ischiorectal fossa (depression between root of tail and sacrotuberous ligament lateral to anus). **27** Hip joint region. **28** Ischiatic tuberosity region (buttock).

Bones of skull
29 Zygomatic (supraorbital) process of frontal bone. **30** External sagittal crest (in dorsal midline of cranium). **31** External occipital protuberance (occiput – most dorsocaudal part of cranium). **32** Nuchal crest (division between dorsal and caudal surface of cranium). **33** Zygomatic arch (bridge of bone connecting face and cranium below eye). **34** Orbit (housing and protecting eyeball). **35** Temporal fossa (origin of temporal muscle). **36** Coronoid process of mandible (insertion of temporal muscle). **37** Mastoid process of temporal bone. **38** Auricular cartilage (basis of pinna of external ear). **39** Scutiform cartilage. **40** Nasal cartilages (surrounding nasal vestibule).

Vertebral column, ribs and sternum
41–46 Cervical (neck) vertebrae. **41** Dorsal arch of atlas vertebra (C1). **42** Wing of atlas vertebra (enlarged flattened transverse process). **43** Transverse foramen of atlas vertebra (passage of vertebral artery). **44** Spinous process of axis vertebra (C2). **45** Transverse processes of cervical vertebrae. **46** 7th (last) cervical vertebra. **47–49** Thoracic

(chest or back) vertebrae. **47** Spinous process of 1st thoracic vertebra. **48** Spinous process of anticlinal vertebra (T10). **49** Spinous process of last thoracic vertebra (T13). **50–51** Lumbar (loin) vertebrae. **50** Transverse process of lumbar vertebrae. **51** Spinous process of lumbar vertebra 7. **52–54** Sacrum (3 fused sacral vertebrae) and caudal vertebrae. **52** Median sacral crest (fused spinous processes of sacral vertebrae). **53** Wing of sacrum (enlarged 1st sacral transverse process). **54** Lateral sacral crest (fused 2nd and 3rd sacral transverse processes). **55** Transverse processes of caudal vertebrae. **56** Cranial articular process of vertebra. **57** Caudal articular process of vertebra. **58** Accessory process of vertebra (on caudal thoracic and lumbar vertebrae). **59** Mammillary process of vertebra (on thoracic and lumbar vertebrae). **60** Intervertebral foramen 1 (lateral vertebral foramen of atlas). **61** Intervertebral foramen 2 (between atlas and axis). **62** Dorsal sacral foramina (passage of dorsal rami of sacral spinal nerves). **63** Lumbosacral junction. **64** Rib 1. Rib **65** 13 (floating rib).

Limb skeleton
66 Dorsal (vertebral) border of scapula. **67** Cranial angle of scapula. **68** Caudal angle of scapula. **69** Spine of scapula. **70** Crest of ilium. **71** Sacral tuberosity of ilium (point of croup). **72** Ischiatic tuberosity (point of buttock). **73** Ischiatic spine. **74** Greater ischiatic notch. **75** Lesser ischiatic notch. **76** Ischiatic arch. **77** Sacroiliac joint. **78** Hip joint. **79** Greater trochanter of femur.

Axial muscles
80 Levator muscle of nostril wing and upper lip. **81** Frontal muscle. **82** Interscutular muscle. **83** Rostral auricular muscles. **84** Caudal auricular muscles. **85** Temporal muscle. **86** Splenius muscle. **87** Lumbar part

of longissimus muscle. **88** Thoracic part of longissimus muscle. **89** Cervical part of longissimus muscle. **90** Thoracic part of iliocostal muscle. **91** Lumbar part of iliocostal muscle. **92** Spinal and semispinal muscle. **93–94** Capital part of semispinal muscle. **93** Complexus muscle. **94** Biventer muscle. **95** Multifidus components of transversospinal muscle. **96** Long tail levators (dorsal medial sacrocaudal muscles). **97** Short tail levators (dorsal lateral sacrocaudal muscles). **98** Lateral tail flexors (caudal intertransverse muscles). **99** Cranial dorsal serrate muscle. **100** Caudal dorsal serrate muscle. **101** External intercostal muscle. **102** External abdominal oblique muscle. **103** Internal abdominal oblique muscle.

Appendicular muscles
104 Latissimus dorsi muscle. **105** Cleidocervical part of brachiocephalic muscle. **106** Cervical part of trapezius muscle. **107** Thoracic part of trapezius muscle. **108** Omotransverse muscle. **109** Thoracic part of rhomboid muscle. **110** Cervical part of rhomboid muscle. **111** Capital part of rhomboid muscle. **112** Infraspinatus muscle. **113** Supraspinatus muscle. **114** Sartorius muscle. **115** Middle gluteal muscle. **116** Superficial gluteal muscle. **117** Deep gluteal muscle. **118** Internal obturator muscle. **119** Gemelli muscles. **120** Iliopsoas muscle. **121** Tensor muscle of lateral femoral fascia. **122–123** 'Hamstring' muscles. **122** Biceps femoris muscle. **123** Semitendinosus muscle.

Ligaments and fascia
124 Sacrotuberous ligament (uniting lateral sacral crest and 1st caudal transverse process with ischiatic tuberosity). **125** Thoracolumbar fascia (dense layer of deep fascia covering epaxial muscles of loins).
(After Ellenberger, Dittrich & Baum, 1956)

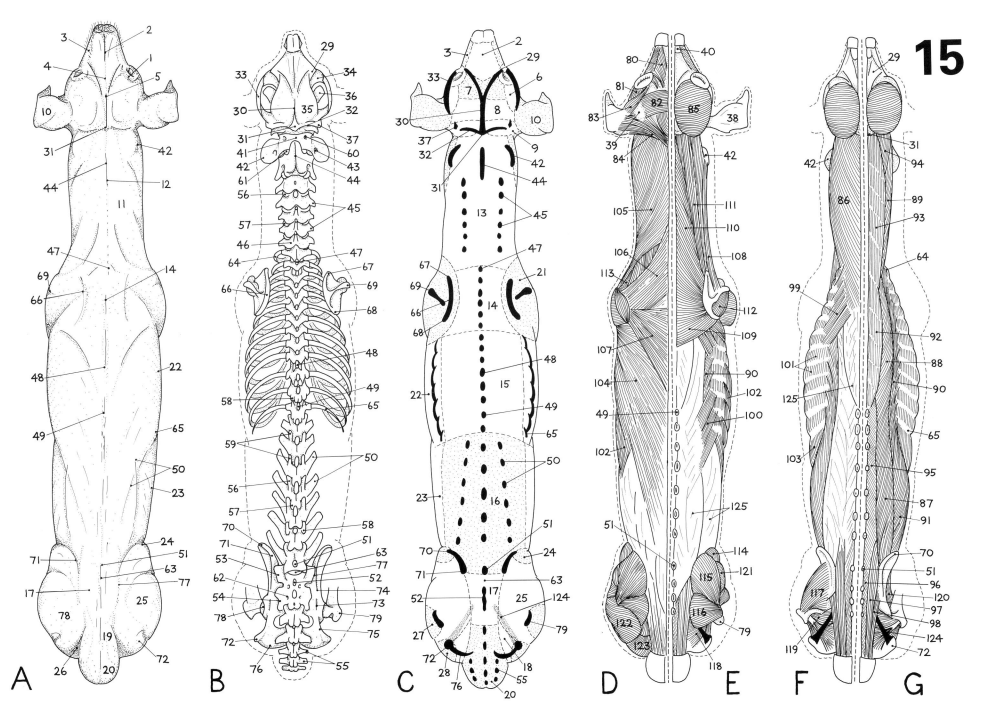

15

A

B

C

D

E

F

G

55

16

SURFACE FEATURES,
SUPERFICIAL MUSCLES AND
SKELETON OF THE DOG
FROM BEHIND

The four illustrations here of the body from the rear show the surface (A), the superficial muscles once the skin has been removed (D), the skeleton (B) and a surface view (C) on which the topographical subdivisions of the body surface are indicated as are the major palpable bony features.

The **ischiorectal fossae** are depressions situated on either side between the root of the tail and the rump muscles roofed over by perineal fascia underlying the skin of the perineum. The lateral wall of a fossa consists of rump muscles, sacrotuberous ligament and ischiatic tuberosity, and the internal obturator muscle on the dorsal surface of the ischium: the medial wall consists of muscles of the pelvic diaphragm (levator ani and coccygeus), external anal sphincter and retractor penis/constrictor vulvae muscles. Ventrally and cranially the fossa is closed off by a merging of fascia on the pelvic diaphragm and internal obturator muscles. From this description I hope you can appreciate that each fossa is a deep wedge-shaped depression lateral to the anus and anal canal and to the terminal end of the urogenital tract in the pelvis. They are normally padded out with fat and loose connective tissue and in older

dogs may be the site of hernia of viscera from the pelvic cavity.

The **hamstring muscles** (*biceps femoris, semimembranosus and semitendinosus*) form the large muscle mass at the rear of the thigh, extending down from the sacrotuberous ligament and ischiatic tuberosity. Unlike the hamstrings in yourself, those of a dog do not terminate in well-defined tendons behind its stifle joint (your own knee joint where you may feel them in yourself quite easily). In a dog they continue down into the lower part of the leg to pass on either side of the calf muscles (gastrocnemius muscles) and attach onto the tibia. Further up in the thigh they have additional attachments onto the femur, patella and patellar tendon, and, as we have already seen, their action on the stifle joint will be both considerable and complex. Extra tendinous cords from the hamstrings (*accessory* or *tarsal tendons*), and also from the gracilis muscle, extend down the rear of the leg where they are closely associated with the tendon from the calf muscles attaching with it on the calcaneal tuberosity. The compound **common calcaneal tendon** attaching to the point of the hock therefore contains: firstly the Achilles' tendon, the tendon of the gastrocnemius muscle beginning midway down the back of the leg; secondly the superficial digital flexor tendon winding around the medial side of Achilles' tendon to spread and cap the point of the hock before continuing down on the underside of the paw to the digits; and thirdly the accessory tendons from the hamstrings and gracilis muscles.

Surface features and topographical regions
1 Sacral region (croup). 2 Root of tail. 3 Caudal region (tail). 4 Gluteal region (rump). 5 Ischiatic tuberosity region (buttock). 6 Clunial region. 7 Ischiorectal fossa. 8 Anal part of perineal region. 9 Urogenital part of perineal region. 10 Root of penis (diverging fibrous structures attached to ischiatic arch). 11 Femoral region (thigh). 12 Popliteal region. 13 Popliteal fossa (at rear of stifle joint). 14 Crural region (crus, shank or lower thigh). 15 Calf. 16 Tarsal region (tarsus, hock or ankle). 17 Calcaneal region. 18 Metatarsal region. 19 Phalangeal region. 20 Metatarsal pad. 21 Digital pad. 22 Scrotum (scrotal region).

Bones, joints and ligaments
23 Median sacral crest (fused sacral spinous processes). 24 Sacral wing (enlarged 1st sacral transverse process). 25 Transverse process of 1st caudal (tail) vertebra. 26 Sacral tuberosity of ilium (point of croup). 27 Wing of ilium. 28 Greater ischiatic notch. 29 Ischiatic spine of hip bone. 30 Lesser ischiatic notch. 31 Ischiatic tuberosity (point of buttock). 32 Ischiatic arch of pelvic girdle (extending transversely between ischiatic tuberosities). 33 Greater trochanter of femur. 34 Lesser trochanter of femur. 35 Trochanteric fossa of femur. 36 Rough surface of femur bounded by medial and lateral lips. 37 Popliteal surface of femur. 38 Lateral condyle of femur. 39 Lateral epicondyle of femur. 40 Medial condyle of femur. 41 Medial epicondyle of femur. 42 Intercondyloid fossa of femur. 43 Lateral supracondyloid tuberosity of femur. 44 Fabellae (stifle sesamoids situated in tendons of origin of gastrocnemius muscle). 45 Lateral condyle of tibia. 46 Medial condyle of tibia. 47 Popliteal notch of tibia. 48 Body (shaft) of tibia. 49 Medial malleolus of tibia. 50 Head of fibula. 51 Body (shaft) of fibula. 52 Lateral

malleolus of fibula. 53 Crural interosseous space between tibia and fibula. 54 Calcaneus (fibular tarsal bone). 55 Calcaneal tuberosity (point of hock). 56 Sustentaculum tali. 57 Talus (tibial tarsal bone). 58 Central tarsal bone. 59 Tarsal bone 4. 60 Tarsal bones 1–3. 61 Metatarsal bones. 62 Proximal plantar sesamoid bones (pair associated with each metatarsophalangeal joint). 63 Hip joint. 64 Stifle joint. 65 Crurotarsal joint. 66 Intertarsal joint. 67 Tarsometatarsal joints. 68 Sacrotuberous ligament (uniting lateral sacral crest and 1st caudal transverse process with ischiatic tuberosity).

Muscles
69 Superficial gluteal muscle. 70–72 Tail muscles. 70 Long tail levators (dorsal medial sacrocaudal muscles). 71 Short tail levators (dorsal lateral sacrocaudal muscles). 72 Lateral tail flexors (dorsal intertransverse caudal muscles). 73 Coccygeus muscle (forming pelvic diaphragm with levator ani muscle). 74 Internal obturator muscle. 75 Ischiourethral muscle. 76 Bulbospongiosus muscle (continuation of urethral muscle around urethral bulb in root of penis). 77 Ischiocavernosus muscle (covering cavernous body in root of penis). 78–80 'Hamstring' muscles. 78 Biceps femoris muscle. 79 Semimembranosus muscle. 80 Semitendinosus muscle. 81 Gracilis muscle. 82 Lateral head of gastrocnemius muscle. 83 Achilles' tendon. 84 Superficial digital flexor muscle and tendons. 85 Deep digital flexor muscle. 86 Common calcaneal tendon (aggregate of structures attaching to point of hock, including Achilles' tendon, superficial digital flexor tendon, and accessory or tarsal tendon of hamstring and gracilis muscles). 87 Interosseous muscles (4 components forming bulk of musculature on plantar surface of metatarsus).

(After Ellenberger, Dittrich & Baum, 1956)

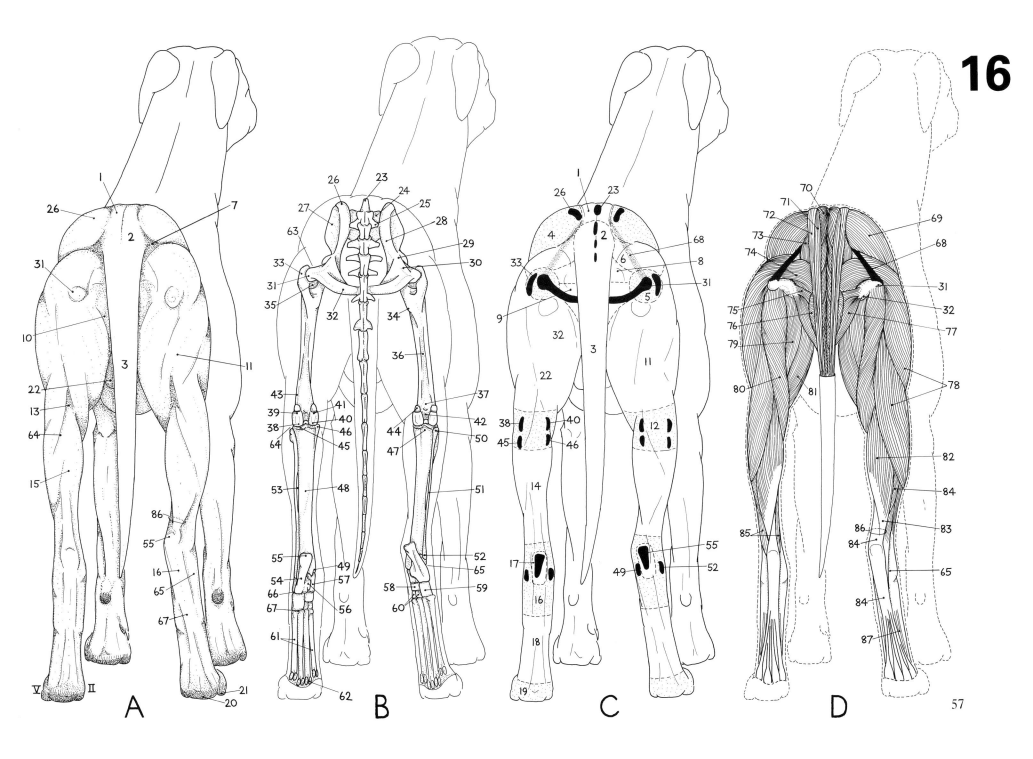

17

SURFACE FEATURES, SUBCUTANEOUS STRUCTURES AND SKELETON OF THE BITCH FROM BELOW

The four accompanying drawings are of a bitch lying on her back. As in previous drawings this sequence shows the surface of the body (A), the skeleton (B) and then subcutaneous structures (D). An additional surface view (C) is shown on which the specifically palpable bony features of the trunk are indicated along with the topographical subdivisions. Such a view is especially useful in attempting to visualize internal structures in relation to surface 'landmarks'.

The surface view shows the typical complement of ten **teats** — two pairs of thoracic, two pairs of abdominal and an inguinal pair. Teats, although not absent in the male dog, are rudimentary and not normally visible through the coat. Teats and their underlying **mammary glands** develop in the young animal along the *milk-line*, a line on either side of the trunk more or less parallel to the midline extending from pectoral to inguinal regions. The glands themselves appear to be highly modified sweat glands of the skin! Since they provide nourishment to a litter of pups they will only be prominent at certain times of a bitch's life — when she is pregnant (also if she is undergoing a pseudo-pregnancy) and throughout the subsequent nursing period. Shrivelling of the milk-producing glandular tissue takes place over a period of a month or two after the pups have been weaned.

Each teat is a short, practically hairless cone of skin which thickens towards its base where a fine covering of hair may be present. Opening onto the surface of a teat are up to a dozen or more small holes each one terminating a teat canal. These canals are the narrowed ends of large spaces inside the gland which receive and temporarily store the milk. Storage of a sufficient amount of milk depends to a considerable extent on the smooth muscle sphincters surrounding teat canals. Reflex relaxation of these sphincters will allow milk to flow when suckling begins. Mammary glands are richly supplied with both blood vessels and lymphatics, the drainage of the latter being of particular interest since it is the pathway taken by malignant cells should they disseminate from mammary tumours particularly in old bitches.

Surface features and topographical regions
1 Nostril region (nasal plane). **2** Oral (mouth) region. **3** Hard palate in roof of mouth. **4** Mental (chin) region. **5** Buccal region (cheek based on buccinator muscle). **6** Masseteric region. **7** Mandibular region (lower jaw). **8** Intermandibular region. **9** Auricular region (pinna of external ear). **10** Outer opening of external ear canal. **11** Lateral neck (jugular) region. **12** Jugular groove. **13** Jugular fossa. **14** Parotid region. **15** Pharyngeal region (throat). **16** Laryngeal region. **17** Tracheal region. **18** Presternal region (breast). **19** Sternal region (brisket). **20** Median pectoral groove. **21** Costal region (thorax, chest or rib region). **22** Cardiac region. **23** Axillary region (including axillary fossa [armpit] between muscles of shoulder and upper arm and muscles of chest wall). **24** Belly. **25–26** Cranial abdominal (epigastric) region. **25** Hypochondriac regions left and right (internal extent of hypochondriac region indicated by dome of diaphragm). **26** Xiphoid region. **27–28** Middle abdominal (mesogastric) region. **27** Lateral abdominal (iliac) regions right and left (flanks). **28** Umbilical region. **29** Fold of flank. **30** Umbilicus (navel). **31–32** Caudal abdominal (hypogastric) region. **31** Inguinal regions right and left. **32** Pubic region. **33** Fold of groin. **34** Teats of mammary glands (cranial and caudal thoracic, cranial and caudal abdominal, inguinal). **35** Urogenital part of perineal region. **36** Vulva (external genitalia, pudendum). **37** Caudal region (tail). **38** Brachial region (arm). **39** Femoral region (thigh). **40** Femoral triangle (bordered by sartorius, pectineus and abdominal muscles).

Bones, joints and ligaments
41 Body of mandible (lower jaw). **42** Angular process of mandible. **43** Upper jaw (supporting teeth of upper dental arch). **44** Internal nostrils (choanae – leading into nasopharynx). **45** Zygomatic arch (bridge of bone connecting face and cranium below eye). **46** Tympanic bulla (surrounding tympanic [middle ear] cavity containing ear ossicles). **47–49** Hyoid apparatus (supporting tongue and larynx). **47** Basihyoid bone. **48** Cranial horn of hyoid. **49** Thyrohyoid bone (caudal horn of hyoid). **50** Thyroid cartilage of larynx. **51** Wing of atlas (C1) vertebra (enlarged transverse process). **52** Enlarged bifid transverse process of cervical vertebra 6. **53** Transverse process of lumbar vertebra 5. **54** Wing of sacrum (enlarged 1st sacral transverse process). **55** Pelvic (ventral) surface of sacrum. **56** Pelvic sacral foramina. **57** Transverse processes of caudal vertebrae. **58** Rib 1. **59** Rib 13 (last or floating rib). **60** Costal arch (fused costal cartilages of ribs 10– 12 attached to costal cartilage of rib 9 last sternal rib – ie. with direct sternal attachment). **61** Costochondral junction between bony rib and costal cartilage. **62–64** Sternum (breastbone). **62** Manubrium of sternum (1st sternebra elongated into base of neck). **63** Sternebrae (sternal segments joined by intersternebral cartilages). **64** Xiphoid cartilage of sternum (last sternebra enlarged into belly wall). **65** Scapula (shoulder blade). **66** Greater tubercle of humerus (point of shoulder). **67** Ilium of hip bone. **68** Coxal tuberosity of ilium (point of haunch). **69** Pubis of hip bone. **70** Pubic pecten. **71** Ischium of hip bone. **72** Ischiatic arch of pelvic girdle. **73** Ischiatic tuberosity (point of buttock). **74** Obturator foramen. **75** Pelvic symphysis (combination of pubic and ischiatic symphyses-fibrocartilaginous). **76** Head of femur. **77** Jaw (temporomandibular) joint. **78** Lumbosacral joint. **79** Sacroiliac joint. **80** Shoulder joint. **81** Hip (coxofemoral) joint.

Muscles
82 Platysma muscle. **83** Sphincter muscle of neck (poorly developed scattering of transversely arranged strands). **84** Cutaneous muscle of trunk. **85** Cleidobrachial part of brachiocephalic muscle. **86** Superficial pectoral muscle. **87** Deep pectoral muscle. **88** Sartorius muscle. **89** Gracilis muscle. **90** Linea alba (fibrous union of lateral abdominal muscles in midventral line of abdomen).

Blood vessels and glands
91 External jugular vein. **92** Mammary glands (cranial and caudal thoracic, cranial and caudal abdominal and inguinal).

Fascial layers
93 Superficial fascia of thigh. **94** Superficial fascia of trunk. **95** Superficial fascia of neck. **96** Axillary fascia.

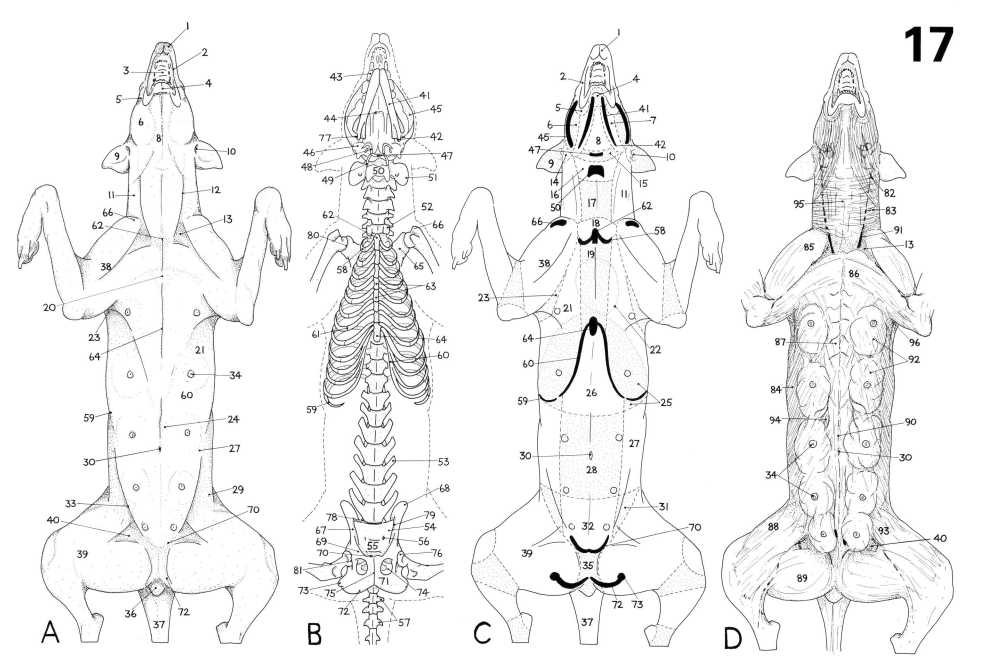

18

SURFACE FEATURES, SUBCUTANEOUS STRUCTURES AND MUSCLES OF THE DOG FROM BELOW

These five drawings of the dog lying on his back show surface and subcutaneous views accompanied by detailed views of the musculature, in particular the muscles of the abdominal and thoracic wall. The topographical subdivision of the abdomen (shown in **A**) into nine regions is based on two transverse and two longitudinal planes. The transverse planes pass through the trunk at the level of the caudalmost extent of the costal arches, and the cranialmost extent of the iliac crests: the longitudinal (sagittal) planes pass on either side of the midline more or less midway between it and the lateral boundary of the flank. These longitudinal lines roughly correspond on the surface to the outer edges of the rectus abdominis muscles (**E**). It should be noted that on the surface the hypochondriac regions (left and right) appear small. However, internally these body regions extend a considerable distance further forwards, in fact as far as the diaphragm this being the cranial boundary of the abdominal cavity internally. Consequently costal and hypochondriac regions 'overlap' quite considerably on the surface. In the male as shown here the preputial and scrotal regions are subregions of the pubic.

This view of the dog also displays the **femoral triangle** on the inside of the upper thigh abutting onto the abdominal wall in the fold of the groin. The base of the triangle is formed by the caudal border of the muscle of the abdominal wall and inguinal ligament. Extending into the thigh the caudal boundary of the triangle is from the **pectineus muscle** a clearly palpable structure arising from the pubic pecten: the cranial boundary is the **sartorius muscle**. The triangle is roofed over by femoral fascia and its floor is lined deeply by the iliopsoas and medial vastus muscles. The principal structures passing through the triangle are the *femoral vessels* and the *saphenous nerve*. These enter the base of a triangle from the abdomen caudal to the inguinal ligament and cranial to the iliopectineal arch (where the deep abdominal fascia blends with the iliac fascia on the iliopsoas muscle). The artery lies cranial to the vein and is an ideal site for taking the dog's pulse.

With the dog on his back the external genitalia are readily visible. The **penis** has a *root* lying below the anus, a *body* extending forwards between the thighs and a *glans* terminating it on the underside of the belly. The divergent *crura* of the penile root arise on either side from the ischiatic arch and are palpable structures where they flank the expanded *penile bulb* immediately below the anus. At the free end of the penis the covering skin forms a **prepuce (sheath)** around the glans. Skin at the preputial opening is continued as a lining layer for the sheath extending back as far as the expanded bulbous part of the glans where it is continued as the surface layer of the glans in turn continuous with the lining layer of the urethra at the external urethral opening. The sheath thus encloses a *preputial cavity* around the glans penis (see especially fig 29C). When a dog attains an erection the body and glans of its penis both enlarge and lengthen, protruding forwards from the sheath which is automatically rolled back onto the enlarged penile body as this occurs. Thus, in addition to providing protection for the sensitive glans penis in the non-erect state, the sheath also provides a reserve skin fold for covering the enlarged body of the penis when erect.

As you will no doubt know the **clitoris** in a bitch is the equivalent structure to the penis. It is a small, all but inconspicuous, structure in the vestibular floor just inside the vulva where it is 'hanging down' below the level of the ischiatic arch (see fig 29F). Like the penis it has root, body and glans, but the body is fatty rather than erectile, and most importantly the urethra does not run through it. However, it does lie in an indentation in the vestibular floor the walls of which are partially folded over the sensitive glans surface comparable to the sheath in a dog. Within the walls of the vestibule some erectile tissue is apparent as **vestibular bulbs**. Traceable into the clitoris these are probably equivalent to the penile bulb of a male which surrounds the urethra in the penile root. This equivalence is understandable since bitch vestibule and dog urethra are developmentally comparable urogenital sinus components. Erectile tissue in the vestibular wall may play a role in forming the 'tie' at copulation locking the penis inside the vestibule and vulva.

The **scrotum** lies about midway between inguinal region and anus, below the body of the penis and between the thighs. This pouch of skin conforms in size and shape to the testes inside it and is covered by thin, pigmented skin which is sparsely haired and liberally supplied with sebaceous glands. Inside the scrotum each testis lies in its own compartment, the two separated by a partition whose position is indicated on the scrotal surface by an indistinct linear marking of the skin. The testes do not lie side by side in the same transverse plane; the left is usually placed further caudally than the right. Remember, the testes are sandwiched between the thighs and this staggered arrangement means that they are subjected to less pressure.

The scrotal skin is underlain by a quantity of fibroelastic tissue containing a well developed layer of smooth muscle. Reflex contraction of this muscle wrinkles scrotal skin bunching up the scrotum and moving the testes closer to the abdomen. This reflex activity presumably assists in regulating testicular temperature since in cold weather the testes are brought closer to the warmth of the abdomen and *vice versa*. Probably the reason why the testes 'descend' to this peripheral position during development is to allow them a slightly lower temperature in which to form sperm. Some anatomical justification for such an assumption is the complete absence of an insulating blanket of subcutaneous fat in the scrotum. However, conclusive proof would seem to be that an undescended testicle retained in the abdomen is unable to produce viable sperm.

In the subcutaneous view (**B**) I have tried to show the cutaneous musculature in the superficial fascia. Of particular note in a dog is the **preputial muscle** passing back into the sheath. Right and left parts pass from the abdominal wall and intermingle in the sheath below the glans attaching to the outer preputial layer. Since the sheath is a structure of some considerable size these muscles prevent its free end from dangling loosely when the penis is non-erect. In a more 'active' sense they may also help roll the sheath forwards over the glans as an erection subsides. An equivalent muscle present in a bitch is nowhere near as well defined, extending back over the abdominal mammary glands. Its functions in the bitch are somewhat difficult to envisage, possibly supporting the teats in some way, or it has been suggested that it might 'erect' teats thereby making them more easily accessible to the suckling pup!

Removal of the cutaneous muscle from the neck (**C**) displays several structures which lie close to the surface, notably the venous tributaries from the head which join up to form the **external jugular vein** passing down the neck to enter the jugular fossa. **Mandibular lymph nodes**, which are normally palpable, and salivary glands are also apparent in this region below the ear and behind the lower jaw.

Musculature of the trunk is shown in various stages of removal in **C**, **D** and **E**. As we have already pointed out abdominal wall muscles form a complete lateral and ventral cover for the abdominal cavity. The musculature is not quite complete being perforated by a pair of openings, the **inguinal canals**,

situated one on either side of the midline at the caudal end just in front of the pelvis in the fold of the groin. The internal opening into a canal leads off from the abdominal cavity while the external opening from a canal leads into the superficial fascia beneath the skin.

Although inguinal canals are present in dogs and bitches their significance is only really apparent in the male. They allow extensions of the peritoneal cavity to pass out of the abdominal cavity (these two cavities are separate structures as I shall explain later). A finger-like outpouching of the peritoneum lining the abdominal wall internally (known as a *vaginal process*) pokes out through each canal and comes to lie in a subcutaneous position below the front of the pelvis (**C**). The peritoneal cavity within the abdomen is continuous with the cavity in each subcutaneous projection. In the dog as shown here the processes pass back through the subcutaneous tissue on either side of the penis below the pelvis, and come together at the caudal surfaces of the thighs below the penile root where they expand inside the scrotum.

The **testes** are contained within these vaginal processes inside the scrotum. Each testis is roughly oval in outline, surrounded by a thick inelastic fibrous coat and is responsible for the continuous production of millions of sperms. Sperm leave the testis and enter a long duct the initial part of which is extensively coiled, embedded in fibrous connective tissue and attached to the surface of the testicle. The whole testicular appendage with its contained duct is called the **epididymis**. At this stage of

their travels spermatozoa are immature and incapable of independent movement. They are passively 'washed out' of the testis by large quantities of testicular fluid. One of the major functions of the duct of the epididymis will therefore be to reabsorb much of this fluid and so concentrate sperm. The enormous length of the duct is correlated with this reabsorptive and storage capacity. Passage of sperm through the epididymal duct takes several days during which they have time to become fully mature. Nevertheless they remain incapable of independent movement and appear to be saving their energy for the long, hazardous ejaculatory journey through the cervix and uterus to the site for fertilization in the uterine tube. The sperm duct eventually leaves the epididymis, straightens out and continues as a muscular tube, the **vas deferens**, which passes forwards inside the vaginal process beneath the skin of the thigh.

In a developing dog, testicles form next to the future kidneys in a sublumbar position at the front of the abdominal roof. Subsequent testicular descent through the inguinal canals inside the vaginal processes enable the testes to attain their definitive position within the scrotum. During its descent in late foetal life a testicle carries its blood vessels, lymphatics and nerves along with it, and also its duct for sperm transport. A testis therefore comes to lie within the scrotum while those structures which it pulled along with it in its descent pass back through the inguinal canal into the abdomen as the **spermatic cord**. In the abdomen testicular vessels pass forwards; an artery coming from the aorta just caudal to the

kidneys, a vein emptying into the caudal vena cava or renal vein close to the kidney. The sperm duct (vas deferens) passes back into the pelvis where its course will be followed later.

Inguinal canals remain open throughout life so that the peritoneal cavity within the abdomen remains in permanent communication with the cavity of the vaginal process inside the scrotum. Because of this state of affairs there always exists the possibility that abdominal organs such as the mobile small intestinal coils might pass into a vaginal sac through an inguinal canal (herniation) especially if the canal were to be enlarged. To reduce the likelihood of this problem occurring an inguinal canal, as its name implies, is not simply a hole, but is an obliquely oriented tunnel. If you think back to the consideration of the abdominal wall you will recall that it consists of three layers of muscle. The entry into an inguinal canal is a penetration through the innermost layer, while the exit from a canal is a penetration through the outermost layer some way further back. The canal's entry and exit are therefore not superimposed on one another, which will in itself reduce the possibility of herniation. In addition the weight of the abdominal organs pressing down on the inside of the belly wall will tend to block off a canal by pressing its internal against its external wall. Nevertheless scrotal hernia is always a possibility in a dog, and inguinal hernia in a pregnant bitch.

Continuing the process of muscle removal in **D** and **E**, the long strap-like muscles lying on either side of the midventral line of the body become apparent. The rectus abdominis muscle extends from attachments on the rib-cage as far forwards as the first rib, back to an attachment on the leading edges of the pubic bones where it merges with a mass of fibrous tissue, the cranial pubic ligament. As the drawing indicates, it has several fibrous intersections in it which suggest that it is a compound structure made up from several smaller segments. Its functions are allied to those of the lateral abdominal muscles considered elsewhere, and there is a very close structural relationship between them. The lateral muscles must pass over the surface of the rectus to reach their attachments on the fibrous tissue of the **linea alba** (white line) in the midline of the underside of the belly. The external and internal oblique muscles pass over its outer (ventral) surface, while the transverse muscle passes over its internal (dorsal) surface (except at the caudal end of the abdomen). Each rectus muscle therefore lies sandwiched between the sheet-like tendons (aponeuroses) of lateral abdominal muscles which form a sheath. On contraction the rectus muscles can move within this sheath.

The midventral fibrous union of the lateral abdominal muscles, the white line, extends from the xiphoid process of the sternum to the prepubic tendon at the pubic symphysis. The **navel (umbilicus)** shown on the skin surface by a whorl of hair and located in the umbilical region, is an area of scar tissue which marks the remains of the point of entry of blood vessels from the developing foetus to the placenta of the mother and *vice versa*. Obviously the scar tissue forms after the umbilical cord is severed,

the vessels shrivel and the umbilicus closes at the birth of a pup. The simple structure of the fibrous linea alba, and the fact that it is not crossed by blood vessels or nerves, makes it a favourable site for surgical incision into the abdomen. However, bearing in mind its capacity for and speed of healing, it cannot compare with that of muscle.

Surface features and topographical regions
1 Mental region (chin). 2 Oral region (mouth). 3 Mandibular region (lower jaw). 4 Intermandibular region (between mandibular bodies). 5 Buccal region (cheek based on buccinator muscle). 6 Masseteric region. 7 Hard palate (in roof of mouth crossed by ridges of cornified buccal mucosa). 8 Auricular region (pinna – visible part of external ear). 9 Parotid region. 10 Pharyngeal region (throat). 11 Laryngeal region. 12 Tracheal region. 13 Lateral neck (jugular) region. 14 Jugular fossa. 15 Presternal region (breast). 16 Sternal region (brisket). 17 Median pectoral groove. 18 Axillary region (armpit between muscles of shoulder and upper arm, and muscles of chest wall). 19 Chest (costal or rib region). 20 Right hypochondriac region. 21 Xiphoid region. 22 Right lateral abdominal region (flank). 23 Fold of flank (extending onto thigh proximal to stifle joint). 24 Umbilicus (navel in umbilical region). 25 Inguinal region (fold of groin between thigh and abdominal wall). 26 Pubic region. 27 Sheath (prepuce covering glans penis). 28 Preputial orifice (leading into preputial cavity). 29 Penis (composed of root, body and glans extending from pelvic outlet forwards between thighs onto underside of belly wall – cut through in figs **D** and **E**). 30 Scrotum (skin sac containing testes and vaginal processes). 31 Scrotal raphe (surface representation of internal subdivision). 32 Root of tail. 33 Thigh (femoral region). 34

Femoral triangle. 35 Arm (brachial region).

Bones, joints and ligaments
36 Auricular cartilage (basis of pinna of external ear). 37 Body of mandible (lower jaw). 38 Zygomatic arch. 39 Basihyoid bone of hyoid apparatus. 40 Thyroid cartilage of larynx (forming laryngeal prominence of voice box). 41 Manubrium of sternum (1st sternebra elongated into base of neck). 42 Xiphoid cartilage of sternum (last sternebra enlarged into belly wall). 43 Rib 1. 44 Costal arch (fused costal cartilages of ribs 10–12 [asternal ribs] attached by fibrous tissue to costal cartilage of rib 9). 45 Rib 13 (last or floating rib). 46 Pubic pecten (supporting a cranial pubic ligament). 47 Ischiatic tuberosity (point of buttock). 48 Ischiatic arch. 49 Os penis (penile bone representing ossified cavernous bodies extending through glans penis).

Muscles and fascia
50 Platysma muscle. 51 Sphincter muscle of neck. 52 Cutaneous muscle of trunk. 53 Preputial muscle. 54 Mylohyoid muscle. 55 Geniohyoid muscle. 56 Digastric muscle. 57 Masseter muscle. 58 Styloglossal muscle. 59 Hyoglossal muscle. 60 Sternohyoid muscle. 61 Sternothyroid muscle. 62 Thyrohyoid muscle. 63 Cricothyroid muscle. 64 Sternocephalic muscle. 65 Scalene muscle. 66 External intercostal muscles. 67 Internal intercostal muscles. 68 Linea alba (white line – fibrous union of left and right sides of belly wall extending from sternum to pubic symphysis). 69 External abdominal oblique muscle. 70 Tendon (aponeurosis) of external abdominal oblique muscle (contributing to external layer of rectus sheath). 71 Internal abdominal oblique muscle. 72 Tendon (aponeurosis) of internal abdominal oblique muscle (contributing to both internal and external layers of rectus sheath). 73 Transverse ab-

dominal muscle. 74 Tendon (aponeurosis) of transverse abdominal muscle (contribution to external layer of rectus sheath). 75 Rectus abdominis muscle. 76 Tendinous inscriptions in rectus abdominis muscle. 77 Aponeurotic tendon of origin of rectus abdominis muscle. 78 Prepubic tendon. 79 Rectus sheath (from tendons of insertion of lateral abdominal muscles). 80 Inguinal canal (through abdominal wall muscles). 81 External inguinal ring (subcutaneous exit from inguinal canal). 82 Internal inguinal ring (abdominal entry into inguinal canal). 83–85 Brachiocephalic muscle. 83 Cleidobrachial part of brachiocephalic muscle. 84 Cleidocervical part of brachiocephalic muscle. 85 Clavicular tendon within brachiocephalic muscle. 86 Superficial pectoral muscle. 87 Deep pectoral muscle. 88 Latissimus dorsi muscle. 89 Teres major muscle. 90 Biceps brachii muscle. 91 Retractor penis muscle (arising from caudal vertebrae and external anal sphincter and travelling along underside of root and body of penis to attach in region of preputial fornix). 92 Cranial and caudal parts of sartorius muscle. 93 Gracilis muscle. 94 Pectineus muscle. 95 Semimembranosus muscle of 'hamstring' group. 96 Adductor muscle. 97 Superficial fascia of thigh. 98 Superficial fascia of trunk. 99 Superficial axillary fascia. 100 Superficial fascia of neck.

Blood vessels and lymph nodes
101 External jugular vein. 102 Maxillary vein. 103 Linguofacial vein. 104 Mandibular lymph nodes.

Glands, viscera and peritoneum
105 Mandibular salivary gland. 106 Parotid salivary gland. 107 Trachea (windpipe). 108 Apex of cranial lobe of right lung. 109 Vaginal process (pocket of abdominal peritoneum extending through inguinal canal into scrotum).

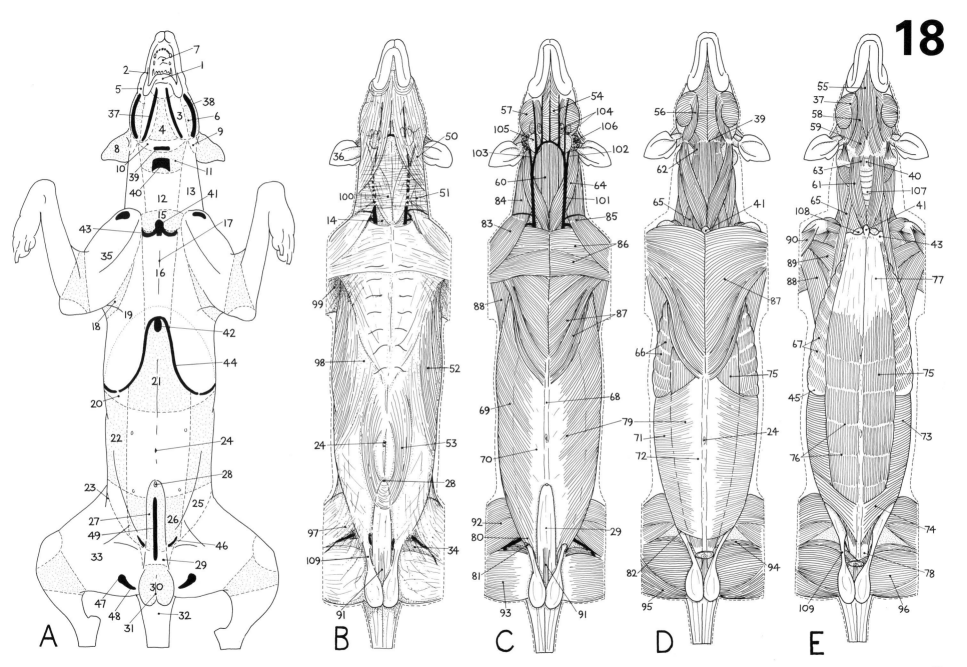

18

63

19

BIOMECHANICS OF THE MUSCULOSKELETAL SYSTEM OF THE DOG

The body of a dog is clearly an integrated system, it cannot be subdivided into separate units independent in their form and function. The preceding drawings have therefore tried to depict the close functional relationship that exists between the muscular and skeletal systems. The text accompanying the drawings has also tried to emphasize their integrated roles in posture and movement. Many of the muscles figured and mentioned in the earlier part of the book are included in the diagrams here but in a somewhat diagrammatic and stylized pattern. Each muscle is depicted simply by lines running along its 'functional axis' between its origin and insertion. In such a format the action of a muscle at any particular joint can be quite readily visualized.

The extrinsic musculature of the limbs is shown in drawings **A** and **D**: the intrinsic limb musculature in drawings **B** and **E**. The epaxial components of the axial trunk muscles are shown in drawing **C**, and the hypaxial components in both drawings **A** and **C**.

Epaxial muscles

1–2 Iliocostal muscle. **1** Lumbar part of iliocostal muscle (*m iliocostalis lumborum*). **2** Thoracic part of iliocostal muscle (*m iliocostalis thoracis*). **3–6** Longissimus muscle. **3** Lumbar part of longissimus muscle (*m longissimus lumborum*). **4** Thoracic part of longissimus muscle (*m longissimus thoracis*). **5** Cervical part of longissimus muscle (*m longissimus cervicis*). **6** Capital part of longissimus muscle (*m longissimus capitis*). **7–9** Semispinal muscle. **7** Thoracic part of semispinal muscle (*m semispinalis thoracis*). **8–9** Capital part of semispinal muscle (*m semispinalis capitis*). **8** Biventer muscle (*m biventer cervicis*). **9** Complexus muscle (*m complexus*). **10** Tail levator muscles (*m sacrocaudalis dorsalis lateralis* and *m sacrocaudalis dorsalis medialis*).

Subvertebral hypaxial muscles

11 Psoas minor muscle (*m psoas minor*). **12** Quadratus lumborum muscle (*m quadratus lumborum*). **13** Longus colli muscle (*m longus colli*). **14** Longus capitis muscle (*m longus capitis*). **15** Tail depressor muscles (*m sacrocaudalis ventralis lateralis* and *m sacrocaudalis ventralis medialis*).

Hypaxial muscles

16 Sternohyoid muscle (*m sternohyoideus*). **17** Sternothyroid muscle (*m sternothyroideus*). **18** Thyrohyoid muscle (*m thyrohyoideus*). **19–20** Sternocephalic muscle (*m sternocephalicus*). **19** Sternooccipital component of sternocephalic muscle (*m sternocephalicus pars occipitalis*). **20** Sternomastoid component of sternocephalic muscle (*m sternocephalicus pars mastoidea*). **21** Scalene muscle (*m scalenus*). **22** External intercostal muscles (*mm intercostales externi*). **23** External abdominal oblique muscle (*m obliquus externus abdominis*). **24** Internal abdominal oblique muscle (*m obliquus internus abdominis*). **25** Transverse abdominal muscle (*m transversus abdominis*). **26** Rectus abdominis muscle (*m rectus abdominis*). **27–28** Muscles of pelvic diaphragm. **27** Coccygeal muscle (*m coccygeus*). **28** Levator ani muscle (*m levator ani*).

Extrinsic muscles of forelimb

29–30 Trapezius muscle (*m trapezius*). **29** Thoracic part of trapezius (*m trapezius pars thoracica*). **30** Cervical part of trapezius (*m trapezius pars cervicalis*). **31** Omotransverse muscle (*m omotransversarius*). **32–33** Rhomboid muscle (*m rhomboideus*). **32** Thoracic part of rhomboid muscle (*m rhomboideus thoracis*). **33** Cervical part of rhomboid muscle (*m rhomboideus cervicis*). **34–35** Ventral serrate muscle (*m serratus ventralis*). **34** Thoracic part of ventral serrate muscle (*m serratus ventralis thoracis*). **35** Cervical part of ventral serrate muscle (*m serratus ventralis cervicis*). **36–38** Brachiocephalic muscle (*m brachiocephalicus*). **36** Brachial part of brachiocephalic muscle (*m cleidobrachialis*). **37** Cervical part of brachiocephalic muscle (*m cleidocephalicus pars cervicalis*). **38** Mastoid part of brachiocephalic muscle (*m cleidocephalicus pars mastoidea*). **39** Latissimus dorsi muscle (*m latissimus dorsi*). **40–41** Pectoral muscles (*mm pectorales*). **40** Superficial pectoral muscle (*m pectoralis superficialis*). **41** Deep pectoral muscle (*m pectoralis profundus*).

Intrinsic muscles of forelimb

42 Supraspinous muscle (*m supraspinatus*). **43** Infraspinous muscle (*m infraspinatus*). **44** Deltoid muscle (*m deltoideus*). **45** Teres major muscle (*m teres major*). **46** Biceps brachii muscle (*m biceps brachii*). **47** Brachial muscle (*m brachialis*). **48** Triceps brachii muscle (*m triceps brachii*). **49** Radial carpal extensor muscle (*m extensor carpi radialis*). **50** Common digital extensor muscle (*m extensor digitorum communis*). **51** Ulnar carpal extensor (*m extensor carpi ulnaris* or *m ulnaris lateralis*). **52** Superficial digital flexor muscle of forelimb (*m flexor digitorum superficialis*). **53** Ulnar carpal flexor muscle (*m flexor carpi ulnaris*). **54** Deep digital flexor muscle of forelimb (*m flexor digitorum profundus*).

Extrinsic muscles of hindlimb

55–56 Iliopsoas muscle (*m iliopsoas*). **55** Psoas major muscle (*m psoas major*). **56** Iliacus muscle (*m iliacus*).

Intrinsic muscles of hindlimb

57–59 Gluteal muscles (*mm glutei*). **57** Superficial gluteal muscle (*m gluteus superficialis*). **58** Middle gluteal muscle (*m gluteus medius*). **59** Deep gluteal muscle (*m gluteus profundus*). **60** Gemelli muscles (*mm gemelli*). **61** Quadratus femoris muscle (*m quadratus femoris*). **62** External obturator muscle (*m obturatorius externus*). **63–66** 'Hamstring muscles'. **63** Biceps femoris muscle (*m biceps femoris*). **64** Semitendinosus muscle (*m semitendinosus*). **65** Semimembranosus muscle (*m semimembranosus*). **66** Tarsal (accessory) tendons of 'hamstring' muscles. **67–68** Quadriceps femoris muscle (*m quadriceps femoris*). **67** Rectus femoris muscle (*m rectus femoris*). **68** Vastus muscles (*mm vasti medialis, intermedius et lateralis*). **69** Patellar tendon. **70** Sartorius muscle (*m sartorius*). **71** Pectineus muscle (*m pectineus*). **72** Gracilis muscle (*m gracilis*). **73** Adductor muscles (*m adductor longus* and *m adductor magnus*). **74** Cranial tibial muscle (*m tibialis cranialis*). **75** Long digital extensor muscle (*m extensor digitorum longus*). **76** Gastrocnemius muscles (*m gastrocnemius caput mediale* and *m gastrocnemius caput laterale*). **77** Superficial digital flexor muscle of hindlimb (*m flexor digitorum superficialis*). **78** Deep digital flexor muscle of hindlimb (*m flexor digitorum profundus*). **79** Linea alba. **80** Sacrotuberous ligament.

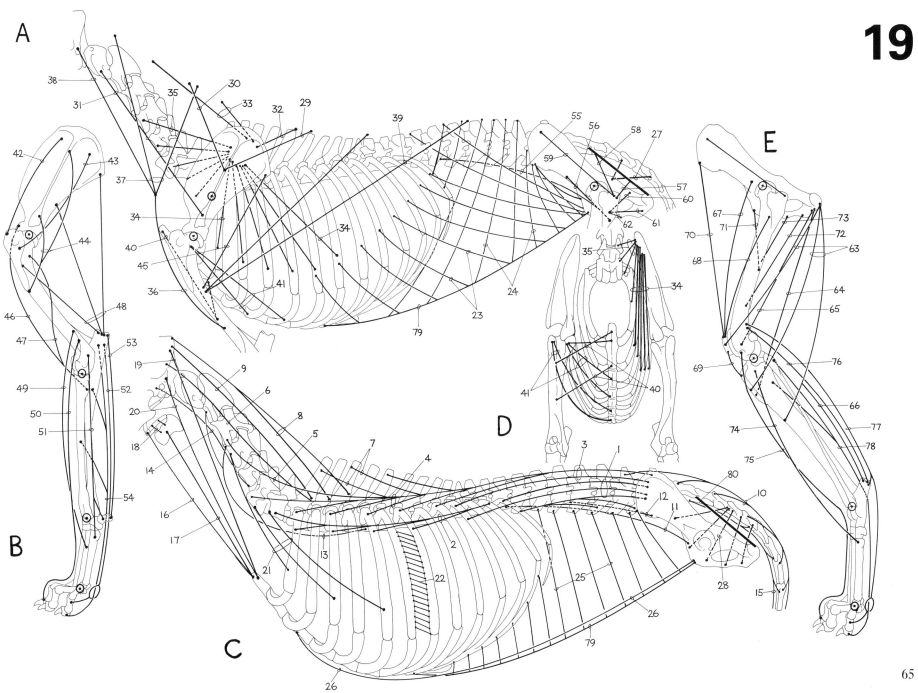

A

B

C

D

E

19

65

20

THORACIC AND ABDOMINAL VISCERA OF THE BITCH FROM THE LEFT SIDE

To begin a sequence of visceral drawings the surface view of the trunk is again shown (**A**) together with a view depicting the topographical regions and palpable reference points to provide us with the 'landmarks' necessary for visualization of the position of internal structures (**B**). The dismemberment that progressively occurred up to fig 14 is continued in **C** − both front and hindlegs have been removed although the hip bone remains in position in the pelvic wall, the hindleg having been removed by disarticulation at the hip joint. The contents of the thorax and abdomen have been exposed by removal of the intercostal muscles in the chest and the transverse abdominal muscle in the flank and belly.

Together with figs 21 & 22 this drawing shows parts of the four organ systems which are prominent in the body cavities of the trunk − digestive, respiratory, excretory and reproductive. Considering first the **digestive system** − as an initial simplification it can be compared to a tube open at both ends (mouth and anus). The cavity inside the gut tube is simply in continuity with the outside world and its lining layer is continuous with the epithelium of the skin at mouth and anus. Unlike skin, however, this epithelial gut lining remains for the most part thin to allow for the absorption of food products and water through it, hence it is potentially fragile. Gut lining is afforded a large measure of protection by a coating of mucus secreted from special epithelial cells. Movement of substances along the tube is brought about by layers of smooth muscle in its walls. Except where modified in the mouth and pharynx, gut muscle is not under a dog's conscious control but acts in a reflex involuntary manner.

The first part of the digestive system, the mouth, concerned with picking up food, tasting it and chewing it, contains the teeth, tongue and related structures all of which will be considered later when the head is dealt with. Assuming that suitable food has been swallowed, it passes into and down the **oesophagus** (gullet) which opens at the back of the pharynx (throat). This is a straightforward transport tube running down the neck alongside the trachea (windpipe). It enters the chest through the thoracic inlet runs back in the roof of the thorax and perforates the diaphragm to enter the abdominal cavity. Waves of muscle contraction move food and fluid along the oesophagus and it is capable of considerable distension allowing for the passage of large pieces of swallowed food. Nowhere along its length is it palpable from the surface even in the neck.

Food passing down the gullet is eventually emptied into the **stomach** for temporary storage. The stomach is tucked up against the left side of the diaphragm; its opening from the oesophagus (the *cardia*) lying in the same plane as thoracic vertebrae 11 or 12; its exit into the small intestine (the *pylorus*) is quite close to the cardia but to the right of the midline. The stomach is therefore orientated more or less directly across the width of the body with a very short concave side (*lesser curvature* joining cardia and pylorus) facing forwards, and a much longer convex side (*greater curvature*) facing backwards. The shape and orientation of the stomach can be readily appreciated if you look at fig 27D. Stomach capacity is variously estimated as between 3 and 8 litres, obviously differing between large and small breeds. Dependent upon the amount of food stored in the stomach, it will also occupy a greater or lesser amount of space within the abdominal cavity of a particular dog. An empty stomach is small and is 'contracted' towards the cardia as a relatively fixed point to the left of the midline. It therefore lies entirely within the 'intrathoracic' part of the abdomen cranial to the costal arch and aligned with ribs 9−12 on the left side. The empty stomach does not contact the belly wall at all being separated from it by the liver and coils of small intestine. A full stomach on the other hand makes extensive contact with the belly wall and may extend back beyond the navel. In **C** the stomach is about half full and appears caudal to the costal arch just in contact with the belly wall caudal to the liver.

As well as storage some initial preparatory digestive mechanisms take place in the stomach. Swallowed food and stomach juices are constantly churned up by regular contractions of its muscular wall to give a thick soupy mass (chyme) which is gradually moved on into the small intestine. Muscular contractions do occur even in an empty stomach − the hunger pangs that I'm sure you have experienced on numerous occasions. The contractions may also be reversed in direction when stomach contents are forced back up the oesophagus as in vomitting. The exit from the stomach has a well developed ring of muscle around it (*pyloric sphincter*) which relaxes regularly allowing stomach contents to be squirted onwards into the small intestine.

The **intestine** (bowel) extends from stomach to anal canal and is broadly divisible into small and large parts. These terms do not refer to the overall dimensions of the two parts (the **small intestine** is in fact considerably longer [approximately 4 m] than the **large intestine** [approximately 70 cm]), but to the size of the cavity inside them. The relatively narrow bore of the small intestine is concerned with food breakdown (digestion) and the subsequent absorption of the broken down products: the wider bore of the large intestine is more concerned with storing waste matter left over after digestion and prior to defaecation, and reabsorbing water from this faecal mass.

The initial part of the small intestine, the **duodenum** is on average 40 cm long and maintains a relatively constant position in the abdominal roof. From the pylorus its 'descending' part passes back in contact with the upper part of the right flank. At the rear of the abdomen, at a transverse plane through lumbar vertebra 5 or 6 ventral to the coxal tuberosity, it curves medially, crosses the midline and passes forwards on the left side of the abdomen forming a U-shaped loop. Its position and shape are quite clearly shown in fig 27D.

The second part of the small intestine the **jejunum/ileum** is considerably longer and is suspended by a more extensive and longer sheet of tissue allowing it a greater range of movement. It appears as a mass of small intestinal coils in contact with the belly wall and lower parts of the flanks. The position of these coils will be shifted backwards when the stomach is filled with food and forwards when the bladder is full of urine. They will also be shifted in the bitch if she is pregnant. The division into jejunum and ileum is somewhat arbitrary since a visible gross distinction is not apparent. In the jejunum, the longer initial part, food passage, digestion and absorption all take place quite quickly so that its cavity usually only contains small quantities of fluid. The ileum on the other hand is normally filled with food because the processes of passage, digestion and absorption are occuring less rapidly.

The large intestine also has two major parts. The short, irregular, blind-ending first part the **caecum**, is approximately 13 cm in length and lies ventral to the descending duodenum about halfway up the right flank in the same transverse plane as the navel. Its position marks the beginning of the large intestine, the ileum joining the large intestine at the junction of caecum and colon. The second part of the large intestine, the **colon**, is another part of the gut which maintains a fairly constant position in the abdominal roof. It begins on the right side in the neighbourhood of the caecum, passes forwards with the descending duodenum along the right flank as the short 'ascending' colon, curves to the left across the midline as

the 'transverse' colon, below thoracic vertebra 12, and then passes caudally on the left side of the abdomen as the 'descending' colon. It roughly corresponds in position to the descending duodenum on the right side. Towards the pelvis it approaches the midline and its shape and disposition are quite clearly shown in fig 27D. The simple continuation of the descending colon through the pelvis is the **rectum**, the line of demarcation being at the pelvic inlet, terminating as the **anal canal**.

In addition to a gut tube the digestive system is also provided with certain accessory organs necessary for the breakdown and assimilation of raw materials. The **liver** is a large, front to back flattened organ divided up by deep indentations into several lobes. It is moulded on the rear surface of the diaphragm maintained in its position at the cranial end of the abdomen by both structural and functional methods. Structurally, ligaments connect liver with diaphragm: functionally, the pressure exerted by other abdominal contents such as the stomach and duodenum effectively sandwich the liver between themselves and the diaphragm. From the surface of the body the furthest cranial extent of the liver is at a transverse plane through intercostal space 7. It is not symmetrically positioned with two thirds of its bulk to the right of the midline. Consequently its right lateral border extends caudally almost to the costal arch corresponding fairly closely with it. On the left side the border of the liver only extends back as far as intercostal space 10. Rarely does it extend caudal to the costal arch except possibly on the right

side and ventrally where it lies on the xiphoid process on the inside of the belly wall in the xiphoid region.

The contribution made by the liver in the digestive process centres around its production of *bile salts*, necessary for fat breakdown and to provide the correct medium in which pancreatic enzymes can function. Bile salts are stored and concentrated in a **gall bladder** before they are passed down a **common bile duct** into the duodenum. The pear-shaped gall bladder lies hidden within a depression in the liver and is therefore not visible in our drawings.

The **pancreas** is a bilobed structure in the abdominal roof. Its right lobe is more or less longitudinally oriented and lies in close contact with the descending duodenum: its left lobe is transverse and lies between stomach and transverse colon. Pancreatic enzymes are required for breaking down many of the compounds present in the diet. They are passed as pancreatic juice in ducts from the pancreas at the region where its two lobes join, into the beginning of the duodenum close to the pylorus of the stomach and in association with the common bile duct.

Stomach contents, well churned up, are squirted through the pylorus into the duodenum to be mixed with bile and with intestinal juices from **duodenal glands**. These additions reduce the acidity of the soupy mass providing a medium in which pancreatic enzymes can bring about digestion. The simpler compounds that food is broken down into are subsequently absorbed through the intestinal wall into minute blood vessels to be transported initially to the liver in the hepatic portal vein.

The large intestine is not primarily a digestive structure. As suggested earlier it reabsorbs water from the waste matter stored inside it gradually forming a firmer faecal mass. Mucus secreted from its walls facilitates movement of the now more solid contents. The large intestine undergoes relatively infrequent periods of activity, unlike the fairly regular churning movements that are carried on in the stomach and small intestine. A few times a day slow pulses of muscular contraction start at a part of the colon swollen with faecal contents and proceed backwards forcing the entire contents along. Once a mass of faeces has been pushed into the descending colon and rectum these parts become sufficiently swollen for a dog to feel the discomfort heralding the need to defaecate.

The **anus** opens at the surface in the perineal region at the pelvic outlet below the root of the tail. It is surrounded by a short *anal canal*, an encircling sphincter of muscle keeping the anus closed except during defaecation. This anal sphincter is actually of two components: an involuntary **internal anal sphincter** which relaxes in a reflex manner as the rectum fills; and a voluntary **external anal sphincter** which a dog can control itself relaxing the sphincter when it is ready to defaecate. Two further involuntary muscles are related to the anal region and cooperate with the anal sphincters to produce what is sometimes termed the 'anal diaphragm'. A **recto-coccygeal muscle** passes back from the rectum onto the underside of the tail beyond the anus: an *anal component of the retractor penis/clitoridis muscle* extends back from the sacrum in the pelvic roof into the rectal wall and anal canal where it blends with the external anal sphincter muscle before continuing to the external genitalia. Together with the coccygeus and levator ani muscles of the pelvic diaphragm, and tail musculature (sacrocaudal muscles), these anal diaphragm muscles contribute to defaecatory movements. The tail is raised by the dorsal sacrocaudal muscles, and the position of the anus is stabilized by the anal part of the retractor penis/clitoridis muscle assisted by the fibrous union of the external anal sphincter with the underside of tail vertebrae 3 and 4. The rectum and anal canal are shortened by rectococcygeal contraction, assisted by raising of the tail which pulls on the rectococcygeus, and squeezed by pelvic diaphragm muscles, assisted by raising of the tail which pulls on these same muscles. The anus is opened on relaxation of the anal sphincter muscles. These defaecatory movements occur practically simultaneously during normal voiding and are accompanied by widespread muscular actions within the trunk which are associated with straining – raising the pressure in the abdomen and pelvis, and with the adoption of the crouched, defaecatory posture.

The anal canal has numerous lubricative mucous glands to assist in faecal passage and also has two large **anal sacs (paranal sinuses)**, one on either side of its opening, sandwiched between internal and external sphincters. These sacs open into the anal canal laterally and contain the foul-smelling secretion from anal glands which is added to the faeces when passed. It may be lubricative assisting defaecation, although its

'scent' presumably had an importance in behaviour patterns such as territorial marking in ancestral dogs, but under domestication it seems to serve little apparent purpose. Indeed, quite often it proves to be a considerable nuisance should the drainage ducts of the sacs become blocked permitting the sacs to swell with accumulated secretion. This will irritate the dog and lead to discomfort. An afflicted dog will try to alleviate its discomfort by chasing its tail or more commonly by squatting and dragging its bottom along the ground.

Surface features and topographical regions
1 Dorsal neck region. **2** Lateral neck (jugular) region. **3** Jugular fossa. **4** Ventral neck (tracheal) region. **5** Presternal region (breast based on superficial pectoral muscles). **6** Sternal region (brisket based on deep pectoral muscles). **7** Scapular region (shoulder). **8** Costal (rib or chest) region. **9** Cardiac region. **10** Axilla (armpit between muscles of shoulder and upper arm, and muscles of chest wall). **11** Left hypochondriac region. **12** Xiphoid region. **13** Left lateral abdominal region (flank). **14** Fold of flank (extending back from belly onto thigh proximal to stifle joint). **15** Umbilical region. **16** Belly. **17** Left inguinal region. **18** Pubic region. **19** Interscapular region (withers). **20** Thoracic vertebral region (back or dorsal region). **21** Lumbar region (loins). **22** Sacral region (croup). **23** Root of tail. **24** Caudal region (tail). **25** Gluteal region (rump). **26** Coxal tuberosity region (haunch). **27** Clunial region. **28** Ischiorectal fossa. **29** Ischiatic tuberosity region (buttock). **30** Shoulder joint region. **31** Brachial region (arm). **32** Cubital region (elbow). **33** Tricipital margin of arm (based on long head of triceps muscle). **34** Olecranon region. **35** Hip joint region. **36** Femoral region (thigh). **37** Cranial margin of

thigh (based on sartorius muscle).

Bones, joints and ligaments
38 Transverse process of cervical vertebra 6. **39** Spinous process of thoracic vertebra 1. **40** Spinous process of lumbar vertebra 1. **41** Spinous process of lumbar vertebra 6. **42** Transverse processes of lumbar vertebrae. **43** Median sacral crest (fused spinous processes of sacral vertebrae). **44** Lateral sacral crest (fused 2nd and 3rd sacral transverse processes). **45** Manubrium of sternum (1st sternebra elongated into base of neck). **46** Rib 1 (bordering thoracic inlet). **47** Rib 6 (denoting approximate caudal extent of heart in chest). **48** Rib 13 (last or floating rib). **49** Costal arch (fused costal cartilages of ribs 10–12 attached to costal cartilage of rib 9 last sternal rib). **50** Dorsal (vertebral) border of scapula. **51** Cranial angle of scapula. **52** Caudal angle of scapula. **53** Spine of scapula. **54** Acromion process of scapula. **55** Greater tuberosity of humerus (point of shoulder). **56** Olecranon process of ulna (point of elbow). **57** Crest of ilium (cranial border of ilium). **58** Coxal tuberosity of ilium (point of haunch). **59** Sacral tuberosity of ilium (point of croup). **60** Ischiatic tuberosity (point of buttock). **61** Ischiatic spine of hip bone. **62** Greater ischiatic notch of hip bone. **63** Lesser ischiatic notch of hip bone. **64** Pubic pecten. **65** Iliopubic eminence. **66** Obturator foramen. **67** Pelvic symphysis. **68** Acetabular fossa of hip bone. **69** Acetabular notch of acetabular fossa.

Muscles
70 Lumbar part of iliocostal muscle. **71** Thoracic part of iliocostal muscle. **72** Lumbar part of longissimus muscle. **73** Thoracic part of longissimus muscle. **74** Cervical part of longissimus muscle. **75** Thoracic spinal and semispinal muscle. **76** Cervical intertransverse muscle. **77** Longus colli muscle. **78** Psoas

major muscle. **79** Tendon of psoas minor muscle. **80** Costal retractor muscle. **81** Costal muscle fibres of diaphragm. **82** Tail levator muscles (dorsal sacrocaudal muscles). **83** Tail depressor muscles (ventral sacrocaudal muscles). **84** Lateral tail flexor muscles (caudal intertransverse muscles). **85–86** Pelvic diaphragm. **85** Coccygeus muscle. **86** Levator ani muscle. **87–88** Anal diaphragm. **87** External anal sphincter muscle. **88** Anal part of retractor clitoridis muscle. **89** Urogenital diaphragm (constrictor muscles of vestibule and vulva). **90** Symphyseal tendon (midline fibrous plate attached to pelvic symphysis and serving for attachment of medial thigh muscles).

Viscera
91 Heart. **92** Trachea (windpipe). **93** Cranial (apical) lobe of left lung. **94** Middle (cardiac) lobe of left lung. **95** Caudal (diaphragmatic) lobe of left lung. **96** Oesophagus (gullet). **97** Stomach (predominantly on left side). **98** Greater omentum (enlarged and modified dorsal mesogastrium associated with stomach but covering much of intestinal mass laterally and ventrally and extensively infiltrated with fat). **99** Descending colon (passing caudally in contact with left flank). **100** Rectum (direct continuation of colon through pelvis). **101** Anus (surrounded by involuntary internal and voluntary external anal sphincter muscles). **102** Liver (exposed slightly beyond costal arch in xiphoid region). **103** Spleen (related to stomach and supported in greater omentum). **104** Left kidney. **105** Left ureter. **106** Urinary bladder (receiving ureters). **107** Left ovary (anchored in sublumbar position by a suspensory ligament). **108** Left uterine (Fallopian) tube (closely related to and encircling ovary). **109** Left uterine horn. **110** Body of uterus (from fusion of horns). **111** Vulva (external genitalia, consisting of vulvar cleft surrounded by labia). **112** Teats of mammary glands.

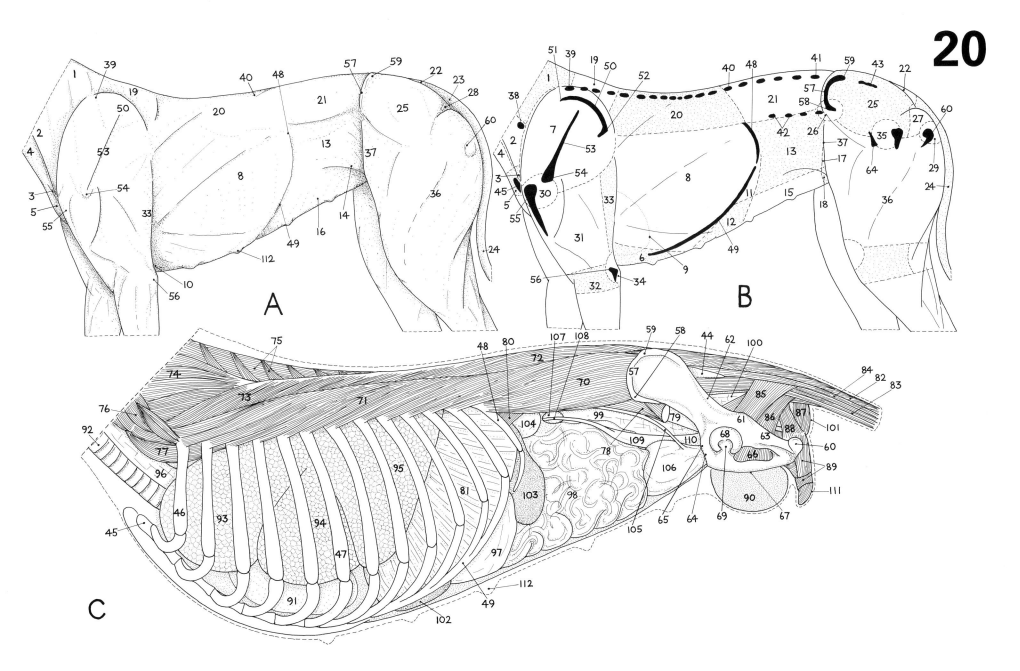

20

A

B

C

21

THORACIC, ABDOMINAL AND PELVIC VISCERA OF THE BITCH FROM THE LEFT SIDE

Exposure of the trunk viscera is continued in these two drawings of a bitch. In **A** a number of ribs have been removed as has the lung exposing the heart and great vessels, and the trachea and oesophagus. Ribs 3 and 6 are left in position now that the heart is exposed since they are 'landmarks' used in mapping its position. The diaphragm, in position in **A**, has had its left half removed in **B**, its cut edge is visible in the midline of the body. This procedure clearly demonstrates the domed shape of the diaphragm, but also exposes those abdominal contents such as the liver and stomach which were tucked up against its rear surface.

In the pelvic wall the hip bone has been removed by disarticulating it at the sacroiliac joint and sawing through the pelvic symphysis. The continuity between abdominal and pelvic cavities is now evident. Removal of the bony pelvic wall leaves the levator ani muscle in position in **A**, the dominant component of the pelvic diaphragm. You may see from this that the levator ani is quite an extensive sheet of muscle arising from both pubis and ilium and flanking the contents of the pelvic cavity. The second component of the diaphragm, the coccygeus muscle, has been removed with the pelvic bone. We have already noticed the importance of the pelvic diaphragm in relation to defaecation and the prevention of possible herniation.

Exposure of the epaxial musculature is also continued in these two drawings. In **A** iliocostal removal exposes the longissimus muscle: in **B** removal of the longissimus exposes elements of the transversospinal system of muscles. With the removal of the lumbar longissimus the continuation into the tail of multifidus components as the sacrocaudal muscles is apparent.

The basic functional plan of the digestive system from oesophagus to anus was outlined with the last drawing and the movements that the stomach and intestines make were mentioned. In order for organs in the abdominal cavity to move about with any degree of efficiency they need to be fairly free of attachments to surrounding structures; ie. they cannot be embedded in 'solid' tissue. Much of the gut and its accessory organs are therefore contained within the abdominal and pelvic cavities and are relatively free to move within them. Organs, however, are not simply floating around freely, they are supported and suspended by sheets of connective tissue (**mesenteries**) dependent from the abdominal and pelvic roof. As well as suspending organs such mesenteries also provide a pathway for blood vessels, lymphatics and nerves to and from them.

In order to reduce friction which would inevitably occur between moving organs some surface lubrication is necessary. Thus the outer surfaces of organs are covered with a fluid secreting (serous) membrane. Likewise the internal surface of the cavity wall is covered by serous membrane of similar nature. These thin, moist, transparent covering layers are the **peritoneal membranes** – *parietal peritoneum* lining the wall is continued onto the surfaces of the mesenteries to become continuous with the *visceral peritoneum* covering the organs. Peritoneum therefore encloses a **peritoneal cavity** common to both abdomen and pelvis in which the viscera are suspended. The peritoneal cavity is opened should the abdominal wall be cut through, although normally (with an intact wall) it is only a potential space, the visceral and parietal peritoneal layers in contact through a film of watery fluid secreted by the peritoneum into the cavity consequently obliterating it. This thin fluid film ensures that movement between surfaces is friction free so that stomach enlargement, gut churning and bladder distension are all permitted. In the pelvis the peritoneal cavity does not fill the available space in the pelvic cavity which means that the terminal part of the rectum and anal canal, and also parts of the urogenital tract in the pelvis, lie *outside* the peritoneal cavity (in a so-called *retroperitoneal* position) embedded in connective tissue and fascia.

The mesentery suspending the gut in the abdomen and pelvis is basically a continuous sheet extending from the diaphragm in front almost to the anus behind. But, this mesentery is not as simple as our description implies since the gut tube which it suspends is considerably longer than the peritoneal cavity containing it. Consequently a complex rearrangement and rotation of both stomach and intestines during foetal development was needed to accommodate a greatly increased length of gut within a peritoneal cavity which is not increasing in linear size at anywhere near the same rate. During gut elongation the intestines rotate around a vertical axis provided by the cranial mesenteric artery: the duodenum becomes hooked around the rear of the artery, while the colon curves around its cranial surface (see fig 27D). This has led to a part of the mesentery becoming twisted around the arterial axis as the *root of the mesentery*; the lengthened mesentery of the duodenum is continuous with it to the left of the axis at the duodenal jejunal flexure, while the lengthened mesentery of the colon is continuous with it to the right of the axis in the neighbourhood of the ileocolic junction close to the caecum. Extending from the root the mesentery fans out to support the coils of the jejunum and ileum.

The mesentery supporting the stomach has also become extensively modified as the **greater omentum**. This structure is shown in position in figs 20 and 22, but has been removed in **A** exposing the coils of the small intestine. In life the greater omentum still runs from the roof of the abdomen to the greater curvature of the stomach, but the stomach has rotated within the abdomen as part of the overall rearrangement of abdominal organs to accommodate them in the peritoneal cavity and now lies practically transverse. The linear arrangement of the mesentery is obviously distorted. As an additional complication the greater omentum no longer runs directly to the stomach but is greatly extended caudally. Thus it descends from the abdominal roof, encloses the left lobe of the pancreas (the right lobe lies in the mesoduodenum), and passes caudally towards the pelvic inlet between the intestines and the ventral belly wall. Just cranial to the pelvic inlet it folds upon itself ventrally and passes forwards to end on the greater curvature of the stomach. If the abdominal wall is incised to open the peritoneal cavity as it would be in many surgical procedures, the two layers of the greater omentum are immediately encountered since they clothe the intestinal coils ventrally (fig 27B) and to a certain extent laterally (figs 20C & 22A). Why the greater omentum has become so enlarged is a difficult question to answer. It does have fat deposited in it and so may serve as a storage depot; while the increased surface area of serous membrane will mean that the capacity for secretion and particularly absorption of peritoneal fluid will be enhanced.

The components of the female reproductive tract are also 'suspended' in the peritoneal cavity. The **ovaries** lie immediately caudal to the kidneys below lumbar vertebrae 3 or 4. The right lies above the descending duodenum; the left contacts the descending colon. Each ovary is held in position by a mesentery thickened as several ligaments, notably a **suspensory ligament**. During oestrus in a bitch (the period of heat which usually occurs twice a year) ovulation takes place and some mature ova are discharged from her ovary into the expanded, funnel-shaped opening of a very narrow, convoluted **uterine tube**. These quite small and delicate tubes are closely associated with the ovaries

supported by the same mesentery and fertilization takes place inside them. Their very narrowness is important in this respect since both ova and sperm are minute structures which could easily 'miss' each other in a large cavity.

The uterine tube abruptly expands as it is continued as the **uterus**, a hollow muscular organ which acts initially as a route by which sperm pass up to reach the ova in the uterine tube, and subsequently provides housing for the developing young, providing attachment for a fertilized ovum passing down from the uterine tube, and as a nourishment source for the developing young by contributing to the formation of a placenta.

In the adult bitch a pair of long and slender **uterine horns** extend back from the uterine tubes to join as a small **uterine body** at about the level of lumbar vertebra 6 or 7. The uterine body passes back in the midline of the abdomen below the colon and above the bladder to terminate as the **uterine cervix** (neck) at the entry into the pelvis. The uterine horns are slightly curved but have a uniform diameter. They are in contact with the sublumbar muscles and ureters in the abdominal roof as well as the descending colon (left horn) and descending duodenum (right horn) further caudally, and lie above the coils of the small intestine and the urinary bladder. These relationships will obviously change during pregnancy. A nonpregnant uterus is very small even in a bitch which has had litters. A pregnant uterus contains a series of enlargements in each horn which represent the position of the developing pups, just like a large string of beads. Each foetus is attached to the **placenta** in the uterine

wall by an umbilical cord which provides a means of transferring material between young and maternal tissue. Normally the developing embryos are evenly distributed between each uterine horn. During the later stages of a pregnancy the heavy and swollen horns sag downwards towards the belly floor, but the ovarian end of the horn maintains a fairly constant position as does the caudal end of the uterus at the pelvic inlet. Thus the enlarged uterine horns bend near their middles as they sink downwards and forwards, in fact they become bent almost double. The pregnant uterus may become sufficiently enlarged so as to contact the stomach the liver and even the diaphragm at the front of the abdomen. Careful palpation in early pregnancy (21 days) may identify the uterine horn with its contained foetuses their number being countable. As the foetuses continue to enlarge they become surrounded by greater quantities of fluid and therefore are more difficult to palpate as individual entities. It is possible that late in a pregnancy (after 50 days) individual foetuses may again be palpable and towards term the foetuses may be felt to move.

The entire uterus (horns and body) is suspended from the abdominal roof, much like the gut is suspended, by sheets of peritoneal mesenteries — **broad ligaments**. These paired ligaments are continuous in front with the mesenteries supporting the ovaries, and each contains some smooth muscle which indicates an active carriage of the pregnant uterus rather than the passive support normally afforded by a mesentery.

The uterine body is very small and

lies at the abdomino/pelvic boundary. The uterine cervix or neck which terminates the body is actually an obliquely oriented canal surrounded by fibrous tissue and some muscle. For most of the bitch's life her cervix is sealed by this ring of muscle. During oestrus, however, it relaxes and opens for a time to allow for sperm entry should the bitch have been mated. It closes again soon after, and during pregnancy a plug of mucus seals it preventing infectious material from entering the uterus. At parturition the cervical muscle relaxes and the cervix dilates to allow passage of the pups. It closes again following birth and only opens up at the next oestrus period.

The **vagina** is a direct continuation back of the uterine body with the uterine cervix projecting somewhat into it. Lying wholly within the pelvis, below the rectum and above the bladder and urethra, this highly dilatable tube passes back to be continued by the vestibule. The 'external' opening of the urethra from the bladder is in the floor of this passage at the vaginovestibular junction. Many female mammals have a hymen in the vaginal opening at this junction; however, in a bitch such a structure is always poorly if at all represented. The **external urethral opening** is therefore some considerable distance from the outer surface of the body at the vulvar cleft, as much as 5 cm in a large bitch (see fig 29F).

The vaginal lining has a different texture to that of the uterus, being thickened and multi-layered, much like the skin. It is also thrown into numerous longitudinal folds. Such a composition is designed to combat the friction its walls are subjected to during copulation and parturition, and allow for the expansion in diameter necessitated during both activities. On either side the vagina is related to the levator ani muscles of the pelvic diaphragm and the ureters cross its lateral surface on their way from kidneys to bladder.

The **vestibule** is the common passageway for genital and urinary systems — a urogenital sinus. Its structure is very similar to the vagina; ie. muscular and dilatable with a stratified, squamous epithelium. It continues back through the pelvis to the vulva, the external opening at the surface of the body. The **vulva** is the only part of the tract which can really be considered as external genitalia and constitutes an opening or *vulval cleft* lying some way below the anus and bounded on either side by thickened lips (*labia*). The area of the body surface surrounding the vulval cleft, the urogenital region (pudendum), is part of the perineum which also includes the anus. At the lower boundary of the cleft the lips join to form a pointed projection hanging down below the level of the ischiatic arch. Labia are soft and pliable, containing fat, elastic tissue, smooth muscle and numerous sebaceous glands. During oestrus (heat) they become enlarged and more prominent. They are in fact the female equivalent of the male scrotum developing from the same areas of the embryo.

Bones
1 13th (last) thoracic vertebra. **2** 7th (last) lumbar vertebra. **3** Sacrum (3 fused sacral vertebrae). **4** Lumbar transverse process. **5** Wing of sacrum (enlarged 1st sacral transverse process). **6** Lateral sacral crest (fused 2nd and 3rd sacral transverse processes). **7** Lumbar accessory process. **8** Rib 1. **9** Rib 3. **10** Rib 6. **11** Rib 13 (last or floating rib). **12** Costal arch (fused costal cartilages of ribs 10–12 attached to costal cartilage of rib 9 [last sternal or true rib]). **13** Manubrium of sternum (1st sternebra elongated into base of neck). **14** Pelvic symphysis (cut through in median plane).

Muscles
15 Longus colli muscle. **16–18** Longissimus muscle. **16** Lumbar part of longissimus muscle. **17** Thoracic part of longissimus muscle. **18** Cervical part of longissimus muscle. **19–21** Semispinal muscle. **19** Thoracic part of semispinal muscle. **20–21** Capital parts of semispinal muscle. **20** Biventer muscle. **21** Complexus muscle. **22–23** Spinal muscle. **22** Thoracic part of spinal muscle. **23** Cervical part of spinal muscle. **24** Multifidus muscle. **25** Cervical intertransverse muscles. **26** Rib levator muscles. **27** Costal retractor muscle. **28–31** Diaphragm. **28** Costal muscle fibres of diaphragm. **29** Central tendinous area of diaphragm. **30** Aortic hiatus of diaphragm (passage of aorta from thorax into abdomen). **31** Oesophageal hiatus of diaphragm (passage of oesophagus into stomach). **32** Sublumbar muscles (cut surface). **33–35** Tail muscles. **33** Tail levators (dorsal sacrocaudal muscles). **34** Tail depressors (ventral sacrocaudal muscles). **35** Lateral tail flexors (caudal intertransverse muscles). **36** Levator ani muscle (major component of pelvic diaphragm). **37** External anal sphincter muscle. **38** Rectococcygeal muscle. **39** Urethral muscle. **40** Anal part of retractor clitoridis muscle (coccygeoanal muscle). **41** Constrictor muscle of vestibule. **42** Constrictor muscle of vulva. **43** Symphyseal tendon (midline fibrous plate attached to pelvic symphysis and providing attachment for adductors and gracilis, the medial thigh muscles).

Heart and blood vessels
44–48 Heart exposed on removal of left lung. **44** Left atrium of heart. **45** Auricular appendage of left atrium of heart. **46** Left ventricle of heart. **47** Right atrium of heart. **48** Right ventricle of heart. **49** Interventricular groove (denoting position of interventricular septum internally). **50** Pulmonary trunk (leading from right ventricle and dividing into right and left pulmonary arteries). **51** Left pulmonary artery. **52** Pulmonary veins (entering left atrium of heart from lungs). **53** Aortic arch (leading off from left ventricle of heart). **54** Thoracic aorta (continuing aortic arch back through thorax). **55** Brachiocephalic trunk (origin of blood vessels to head and right forelimb). **56** Left common carotid artery. **57** Left subclavian artery (to left forelimb — arising directly from aortic arch). **58** Ligamentum arteriosum (occluded ligamentous representation of a patent foetal communication between aorta and pulmonary trunk, ductus arteriosus, which bypassed lungs in foetal life. Occlusion occurs at birth when lungs expand at first breath).

Lymph nodes
59 Caudal deep cervical lymph node (only member of deep cervical nodes consistently present). **60** Cranial mediastinal lymph node (one of up to 6 nodes). **61** Middle tracheobronchial lymph node (large node at tracheal bifurcation). **62** Sternal lymph node.

Autonomic nerves
63 Left vagosympathetic trunk (association of cervical sympathetic trunk and vagus nerve in neck). **64** Left vagus nerve. **65** Left recurrent nerve (hooked around aortic arch prior to ascent of neck on trachea). **66** Dorsal vagal trunk (joins dorsal vagal trunk of right side on oesophagus). **67** Ventral vagal trunk (joins ventral vagal trunk of right side on oesophagus). **68** Cardiovagal nerve.

Internal organs
69 Thymus gland (shown here as the size it might be in a juvenile — in later life regresses considerably although always present). **70** Trachea (windpipe). **71** Left principal bronchus (from tracheal bifurcation). **72** Cranial (apical) lobe of right lung. **73** Caudal (diaphragmatic) lobe of right lung. **74** Intermediate (accessory) lobe of right lung. **75** Oesophagus (gullet). **76–78** Stomach (predominantly on left side of body). **76** Fundus of stomach (bulging dorsally). **77** Body of stomach (orientated transversely). **78** Greater curvature of stomach (facing caudoventrally). **79** Small intestine (numerous coils of jejunum and ileum occupying much of abdominal cavity and removed in B). **80** Descending colon (passing caudally against left flank). **81** Rectum (continuation of colon through pelvis) **82** Anus (surrounded by internal and external anal sphincter muscles). **83–84** Liver (moulded on rear face of diaphragm). **83** Left lateral lobe of liver. **84** Left medial lobe of liver. **85** Spleen (related to stomach and supported in greater omentum). **86** Left kidney. **87** Left ureter. **88** Urinary bladder (receiving ureters from kidneys). **89** Urethra (continuation of neck of bladder). **90** Left ovary. **91** Left uterine (Fallopian) tube (closely related to ovary). **92** Left uterine horn. **93** Body of uterus (formed from fusion of uterine horns). **94** Position of cervix of uterus (sphincteric neck terminating uterine body). **95** Vagina (continuing cervix into pelvic cavity). **96** Vestibule (urogenital sinus, direct caudal continuation of vagina, entry of urethra marking vaginovestibular junction). **97** Vulva (external genitalia — consisting of vulvar cleft surrounded by labia).

Body cavities
98 Thoracic cavity. **99** Abdominal cavity. **100** Pelvic cavity (in continuity with abdominal cavity through pelvic inlet).

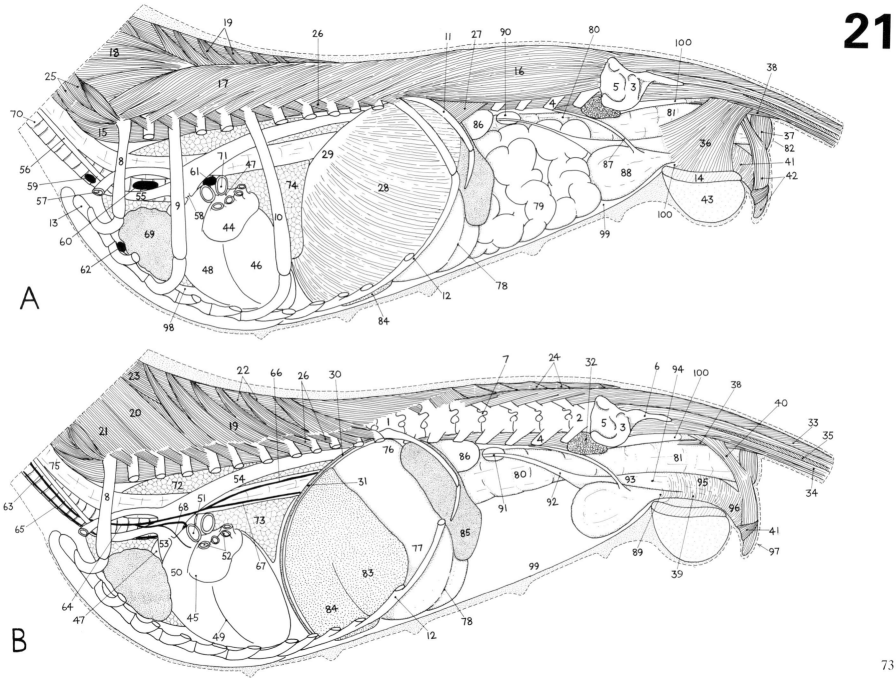

A

B

22

THORACIC, ABDOMINAL AND PELVIC VISCERA OF THE DOG FROM THE RIGHT SIDE

These two drawings of the male are very much like previous figures of the bitch except that we are viewing him from the right side. The upper drawing (**A**) is directly comparable with fig 20C being at about the same stage of dissection (except for the axial musculature). In **B** removal of ribs and the right lung exposes the heart and great vessels from the right side, especially those vessels entering the right atrium of the heart. In the abdomen removal of the greater omentum and the right half of the diaphragm exposes the abdominal contents, and you will notice the greater bulk of liver on this right side.

The final few components of the axial musculature are displayed here. In **A**, following removal of the spinal and semispinal components and the dorsal sacrocaudal muscles, the multifidus components of the transversospinal system are clearly displayed as short, discrete entities. The supraspinous ligament and its continuation into the neck, the nuchal ligament, are also shown. In **B** practically all of the remaining axial muscles have been removed except for the short rib rotator muscles and the interspinal muscles in the trunk.

Some elements of the **respiratory system** are shown in these drawings as well as in figs 20 & 21. As an initial simplification it can be compared to a pair of sacs inside the body open to the outside at one end only (unlike the digestive tube already considered). The epithelial lining of much of the tube is a mucus producing membrane continuous with the epithelium of the skin at the external nostrils. Like the gut the respiratory tube has its own reflex or involuntary visceral muscles. However, they are nowhere near so extensive since breathing movements are brought about primarily by general body muscles of the skeletal or voluntary type rather than by visceral muscles which in the respiratory system only regulate airflow.

Respiratory air is drawn in from the body surface through a series of passageways, the *upper respiratory tract*, to the lungs where an exchange of gases takes place – oxygen is taken in and carbon dioxide is passed from the lungs back along the passages of the upper respiratory tract to the outside. The respiratory system must also therefore incorporate a mechanical apparatus for producing the airflow - pleural cavities, ribcage and respiratory muscles.

The upper respiratory tract in the head beginning at the external nostrils will be dealt with a little later. Assuming that inspired air has been drawn through the nose and pharynx it enters the **trachea** (windpipe) through the **larynx**, a valve guarding its entrance. This flexible tube extends from the larynx at the back of the throat down the neck in close association with the oesophagus and through the thoracic inlet into the chest. It ends by dividing into *left and right principal bronchi* somewhere below thoracic vertebra 4. The trachea must be resistant to collapse since it only contains air which would be unable to open a collapsed tracheal cavity. It must also be capable of distortion when the head and neck are moved and when chunks of food are swallowed and pass down the neighbouring oesophagus. These two provisos make it imperative that it have a flexible supporting framework. The trachea therefore has a 'skeleton' of approximately 35 separate ring-shaped **tracheal cartilages** which prevent collapse whilst still allowing flexibility.

Within the chest each principal bronchus branches and rebranches to give a '*bronchial tree*' of air passages permeating throughout the lungs and in fact providing the scaffolding upon which they are built. In the terminal branches of the tree (the *lower respiratory tract*) air tubes are termed bronchioles and they expand into minute thin-walled cavities (air sacs or pulmonary alveoli) which represent the final subdivisions of the airways. Gaseous interchange occurs between the air in an alveolus and blood cells and plasma in blood capillaries in the alveolar walls. At a microscopical level the functional division between air conduction and gas exchange areas within the bronchial tree is shown by a change in the nature of its epithelial lining, and by a loss of fibrous and cartilaginous tissue in its walls. The epithelium of the conduction pathway through trachea and bronchial tree is ciliated and both serous and mucus secreting. Cilia are minute, hair-like projections from the free surface of the cells that can beat rhythmically to move mucus over the cells' surface. This is obviously of advantage to a dog since particulate matter entangled in the mucus can be moved up the trachea away from the lungs by ciliary activity. In the gas exchange region the epithelium lacks both cilia and mucus, the cells only secreting a thin film of fluid.

Bronchi, bronchioles, alveoli and the pulmonary blood vessels and nerves to the bronchial tree are all joined together by connective tissue to produce the overall shape and form of a lung. **Lungs** have a spongy texture because of their permeating network of air-filled spaces, and occupy a considerable portion of the available space within the chest. Each lung is shaped somewhere between a cone and a pyramid; the bases facing back, the apices at the thoracic inlet, although the right lung is larger than the left. Deep indentations (fissures) also subdivide the lungs into several lobes.

Lungs along with several other structures are contained within the **thoracic (chest) cavity**. The cavity is more or less cone-shaped with a blunt apex at the thoracic inlet and an oblique concave base at the thoracic outlet (blocked by the diaphragm). Unlike the abdominal cavity, however, the thoracic cavity is also completely divided into right and left parts by a vertical partition of connective tissue, the **mediastinum**, extending from thoracic inlet to diaphragm. The mediastinum separates right from left lung and contains *all* of the remaining thoracic organs – trachea, oesophagus, heart and great vessels – packed around by connective tissue. At the tracheal division left and right principal bronchi pass out of the mediastinum on either side to enter the lungs. Along with a principal bronchus run pulmonary blood vessels, nerves and lymphatics, the aggregate of structures forming the *root of the lung*, anchoring it to the mediastinum.

The **diaphragm** is the partition between thorax and abdomen and consists of radially disposed muscle fibres converging onto a *central tendinous* area. The muscle tissue originates from the inner surfaces of the costal arches, the xiphoid cartilage of the sternum and from the underside of lumbar vertebrae 3 and 4. The marked cranial convexity of the diaphragm extends forwards from these attachments into the ribcage and its most cranial extent may reach a transverse plane through ribs 6 or 7. Internally this brings it very close to the heart and there is in fact a ligamentous connection between the diaphragm and the pericardium around the heart. During quiet breathing when at rest the central tendon moves very little so that the position of the heart varies only slightly during the respiratory cycle.

The lumbar origin of the diaphragm is in the form of two thickened bands of muscle (crura) which join below the main aorta and give rise to fibres fanning out into the tendinous centre (fig 27E). The **aortic opening (hiatus)** in the diaphragm is in the dorsal midline below thoracic vertebra 13 and is actually surrounded by a tendinous aortic ring necessary to prevent aortic constriction on diaphragmatic contraction. The diaphragm is also perforated by the oesophagus and this **oesophageal hiatus** is through the crural muscle just to the left of the midline below thoracic vertebra 11. As with the aortic hiatus the muscular margin of the oesophageal opening is not firmly attached to the oesophagus. The muscle loosely encircles it providing no impediment to swallowing, enabling the crus to act as a sphincter, but at the same time does make this the weakest part of the diaphragm. The caudal vena cava also penetrates through the diaphragm, but the **caval foramen** is through the central tendon to the right of the midline in the transverse plane of thoracic vertebra 8 or 9. This hiatus is different in that the connective tissue in the vessel wall is actually fused with the central tendon completely sealing the foramen around the vessel. Diaphragmatic hernias are encountered either through actual tears in the muscular or tendinous parts, or possibly through the oesophageal hiatus.

Just like abdominal viscera lungs must be able to move without friction at their surfaces. They must also be capable of expansion and contraction. Both needs are met by a serous membrane lined **pleural cavity** associated with each lung. On either side of the chest the inside of the wall, the cranial surface of the diaphragm and the lateral face of the mediastinum are covered by a layer of fluid producing membrane (*parietal pleura*) continuous at the root of the lung with a pleural layer covering the surface of a lung (*visceral or pulmonary pleura*). A minute amount of pleural fluid produced by the pleurae, occupies the pleural cavities as a fluid film which maintains the contact between visceral and parietal pleurae whilst still allowing the layers to slide over one another. Normally the pleural cavities are therefore only potential and not actual cavities.

In addition to separating thoracic and abdominal cavities the diaphragm is the dominant muscle responsible for producing an airflow. On contraction it increases the length of the thoracic cavity by a flattening of the muscular periphery of its dome (there is little caudal movement of the central tendon). Rib movements increasing the width of the thorax, accomplished by intercostal muscles (as explained earlier), are only of importance in deeper levels of respiration.

Breathing movements are based on regular changes in size of the thoracic cavity. Such changes will automatically involve changes in size of the pleural cavities contained within the thorax since the parietal pleura is attached to the internal surface of the thoracic wall. In turn the lungs will passively follow these movements. Thus on chest expansion the lungs will expand and the air pressure within them will be lowered. As long as the pressure falls below that of the surrounding atmosphere, air will be drawn in through the upper respiratory tract. For the process to occur at all the parietal and visceral pleural layers must be intact and adherent through the thin film of fluid produced by them. If anything should happen to disrupt this fluid film then visceral may pull away from parietal pleura and the potential cavity will become an actual one. Lung collapse will occur in such a situation because healthy lungs contain a large amount of elastic fibres which are under constant tension. In fact most of the muscular effort involved in breathing in is concerned with stretching this tissue even further. Hence breathing out is essentially passive, the natural elasticity of a stretched lung will decrease its size and pull the ribs and diaphragm back into position once the inspiratory muscles have relaxed, thus expelling air. However, should an increased rate of breathing be required to allow for increased activity then passive expulsion of air by elastic recoil of lung tissue may not be rapid enough and an active expiratory effort will be needed. This is provided by the abdominal muscles whose contraction raises intraabdominal pressure and pushes abdominal contents (stomach, liver, etc) against the rear of the diaphragm. This is pushed forwards against the lungs compressing them and actively pumping air out of them.

Since these two drawings depict a dog the pelvis differs considerably from the preceding figure of a bitch. In **B** the pelvic cavity has been similarly opened by removing the hip bone and pelvic diaphragm from its lateral wall. The components of the male urogenital tract continuing back through the pelvis to exit from its rear, the pelvic outlet, and enter the root of the penis are clearly displayed. The testis, epididymis and vas deferens are somewhat diagrammatically displayed. The sperm duct passes within the vaginal process across the lateral surface of the penile body beneath the skin to enter the peritoneal cavity in the abdomen through the inguinal canal. Inside the abdomen the **vas deferens** inclines caudally, loops over the ureter leading to the urinary bladder and enters the pelvis. At this point it converges with its fellow of the opposite side above the bladder neck and the two ducts empty together into the **urethra**. Sperm movement along the vas deferens occurs immediately prior to ejaculation and results from muscular waves passing along the walls of the duct – sperm at this stage are *still* in effect non-motile.

Beyond the point of entry of the sperm ducts the **urethra** is a commom passage for both sperm and urine, but the flows obviously occur at different times. It continues through the pelvis and penis opening at the tip of the glans. As you may see from the drawing (**B**) the point of entry of sperm ducts into urethra is surrounded by the **prostate gland**. The ejaculate (semen) is a combination of sperm from the testes and fluid from the prostate emptied into the urethra through numerous minute openings. It is somewhat difficult to suggest what prostatic fluid actually does although it has been suggested that it provides an energy source for the sperm because it contains some simple sugars. It might also alter the composition of the fluid in which the sperm is carried. In the epididymis fluid becomes increasingly acid because sperm concentration leads to a build up of carbon dioxide produced by their respiratory activity. Prostatic secretion may neutralize this acidity producing the required 'environment' in which sperm can develop to the full their capacity for independent mobility. A further possible function for prostatic fluid will be mentioned later when the mechanics of copulation are considered.

Bones, joints and ligaments
1 Transverse process of thoracic vertebra 4. **2** 13th (last) thoracic vertebra. **3** Transverse process of lumbar vertebra 4. **4** 7th (last) lumbar vertebra. **5** Median sacral crest (fused sacral spinous processes). **6** Wing of sacrum (enlarged 1st sacral transverse process). **7** Lateral sacral crest (fused 2nd and 3rd sacral transverse processes). **8** Caudal vertebra 1.

9 Accessory processes on caudal thoracic and lumbar vertebrae. **10** Mammillary processes on thoracic and lumbar vertebrae. **11** Intervertebral foramen (for passage of spinal nerves and vessels). **12** Supraspinous ligament. **13** Nuchal ligament. **14** Rib 1. **15** Rib 3. **16** Rib 6. **17** Rib 13 (last or floating rib). **18** Costal arch (fused costal cartilages of ribs 10-12 [asternal or false ribs] associated with costal cartilage of rib 9 [last sternal or true rib]). **19** Manubrium of sternum (1st sternebra elongated into base of neck). **20** Wing of ilium of hip bone. **21** Crest of ilium. **22** Coxal tuberosity of ilium (point of haunch). **23** Sacral tuberosity of ilium (point of croup). **24** Ischium of hip bone. **25** Ischiatic spine. **26** Greater ischiatic notch. **27** Lesser ischiatic notch. **28** Ischiatic tuberosity (point of buttock). **29** Pubis of hip bone. **30** Pubic pecten. **31** Obturator foramen. **32** Acetabular fossa of hip bone. **33** Pelvic symphysis.

Muscles
34 Symphyseal tendon (fibrous midline plate attached to pelvic symphysis and serving for origin of medial thigh muscles). **35** Sublumbar muscles. **36** Longus colli muscle. **37–41** Transversospinal muscles. **37** Lumbar part of multifidus muscle. **38** Thoracic part of multifidus muscle. **39** Rib levator muscles. **40** Rotator muscles. **41** Interspinal muscles. **42–44** Diaphragm. **42** Costal muscle fibres of diaphragm. **43** Central tendinous area of diaphragm. **44** Caval foramen of diaphragm (passage of caudal vena cava). **45** Tail depressors (ventral sacrocaudal muscles). **46** Tail levators (dorsal sacrocaudal muscles). **47** Lateral tail flexors (caudal intertransverse muscles). **48** Diagrammatic representation of position of inguinal canal. **49–50** Pelvic diaphragm. **49** Coccygeus muscle. **50** Levator ani muscle. **51–54** Urogenital diaphragm. **51** Urethral muscle (around pelvic urethra). **52**

Bulbospongiosus muscle (covering surface of penile bulb). **53** Retractor penis muscle. **54** Ischiocavernosus muscle (on root of penis attached to ischiatic arch). **55–57** Anal diaphragm. **55** External anal sphincter muscle. **56** Rectococcygeal muscle. **57** Anal part of retractor penis muscle.

Heart and blood vessels
58–59 Heart. **58** Right atrium of heart. **59** Right ventricle of heart. **60** Coronary groove of heart. **61** Caudal vena cava (returning blood from rear end of body to heart). **62** Cranial vena cava (returning blood from front end of body to heart). **63** External jugular vein. **64** Internal thoracic (mammary) vein. **65** Subclavian vein (draining blood from forelimb). **66** Costocervical vertebral trunk. **67** Azygos vein.

Autonomic nerves
68 Vagosympathetic trunk (association of vagus nerve and cervical sympathetic trunk). **69** Vagus nerve. **70** Dorsal vagal trunk. **71** Ventral vagal trunk. **72** Recurrent nerve. **73** Subclavian loop of sympathetic trunk (subdivision of trunk around subclavian artery). **74** Cervicothoracic (stellate) sympathetic ganglion (fused last cervical and first three thoracic sympathetic trunk ganglia). **75** Sympathetic trunk. **76** Vertebral nerve (transverse nerve – combined grey communicating rami to cervical nerves).

Internal organs
77 Trachea (windpipe). **78–81** Right lung. **78** Cranial (apical) lobe of right lung. **79** Middle (cardiac) lobe of right lung. **80** Caudal (diaphragmatic) lobe of right lung. **81** Cardiac notch in ventral border of right lung. **82** Left lung (partly visible). **83** Principal bronchus of right side (arising from tracheal bifurcation). **84** Oesophagus (gullet). **85** Stomach (only

just visible on right side). **86** Descending duodenum (passing back against right flank). **87** Pancreas. **88** Small intestine (coils of jejunum and ileum in abdominal floor). **89** Greater omentum (enlarged and modified dorsal mesogastrium). **90** Caecum (in centre of right flank). **91** Descending colon. **92** Rectum (continuing colon through pelvis). **93** Anus. **94** Liver. **95** Right kidney. **96** Right ureter. **97** Urinary bladder (receiving ureters from kidneys). **98** Prostate gland (surrounding pelvic urethra at pelvic inlet). **99** Urethra (continuation of neck of bladder). **100** Penile (urethral) bulb (at root of penis). **101** Right crus of penis (attached to ischiatic arch medial to ischiatic tuberosity – cut through after removal of hip bone). **102** Body of penis (commences at merging of penile crura). **103–104** Glans penis. **103** Bulbus glandis (expansion of erectile tissue at base of glans penis). **104** Pars longa glandis (bulk of glans penis containing os penis). **105** External urethral orifice (at free end of glans penis). **106** Scrotum (containing testes). **107** Testis. **108** Epididymis (appendage of testis containing epididymal duct in which sperm is stored and concentrated). **109** Vaginal process of parietal peritoneum. **110** Spermatic cord (composed of testicular blood vessels and vas deferens located within vaginal process). **111** Vas deferens (conveying sperm away from testis). **112** Sheath (prepuce). **113** Preputial fornix (at which sheath reflects onto glans penis). **114** Preputial cavity (around glans penis). **115** Preputial orifice (at tip of penis and leading into preputial cavity).

Body cavities
116 Thoracic cavity (separated from abdominal cavity by diaphragm). **117** Abdominal cavity. **118** Pelvic cavity (in continuity with abdominal cavity through pelvic inlet).

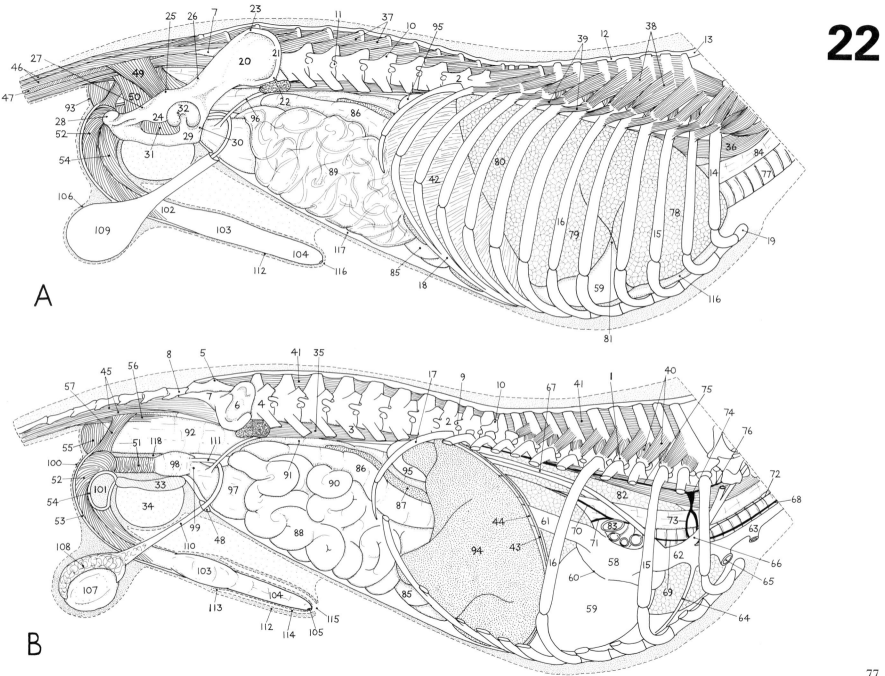

22

A

B

77

23

SURFACE PROJECTIONS OF THORACIC, ABDOMINAL AND PELVIC VISCERA OF THE DOG FROM THE LEFT SIDE

These drawings are simple sketches of the body on which the outlines of various internal structures are superimposed to indicate their approximate position in relation to the surface. You must remember that these positions are to some extent arbitrary since a number of factors will modify the position and form of viscera. Obviously posture will influence visceral position – whether the dog is standing up or lying down on its back or on its side. Respiratory movements will influence the position of thoracic but also of abdominal contents, with deep inspiration displacing the viscera caudally. However, visceral distension in the abdomen will be the most important modifier of position. This applies especially to the stomach and bladder, but also to the spleen, and the uterus in various stages of pregnancy.

Attempting to 'visualize' the position of internal organs is necessary since little of them can be identified or palpated from the surface. This is true for thoracic viscera and the abdominal viscera lying within the area bounded by the caudal ribs, costal arch and diaphragm. The 'intrathoracic' abdominal contents include the liver, stomach, part of the spleen and kidneys, and parts of the small and large intestines. Caudal to the costal arch the remainder of the abdomen in the male dog is occupied predominantly by the intestinal mass. But even here tone in abdominal wall muscles offers resistance to palpation and therefore identification. Nevertheless a number of viscera occupy fairly constant positions. In order to simplify matters and to assist you in visualizing position and relationships the thorax and abdomen are dealt with in separate pictures.

As a starting point for our consideration I have again included at the top of the page (**A**) the lateral view of the trunk on which a number of palpable bony points of reference are indicated along with the topographical subdivisions of the body. We are becoming quite familiar with this drawing and I'm sure that you can now name most of the components without having recourse to the accompanying legend!

The bony limits of the **thoracic cavity** are indicated in the thorax drawing (**B**) by the continuous line. Its roof is formed from thoracic vertebrae underlain by longus colli muscles; its walls from ribs and costal cartilages and the intervening intercostal muscles; its floor from the sternum overlain by transverse thoracic muscles; its cranial boundary is the thoracic inlet between the first ribs; its caudal boundary is the thoracic outlet closed by the diaphragm. You can see that the cavity is smaller than the surface contour of the thorax especially dorsally where the upper contour of the back is some distance away from the dorsal boundary of the thoracic cavity. *The overall depth of the thorax is therefore considerably in excess of the actual depth of the cavity inside it.* The cavity is also restricted in width as well as depth at the thoracic inlet where it lies between the upper ends of the limbs (fig 13B). In cross section the thorax is somewhat oval although wider dorsally and narrower ventrally, but the cavity is more or less 'heart-shaped' since on either side of the line of the vertebral bodies the cavity extends further dorsally because of the upward arching of the ribs.

The *apparent* extent of the thoracic cavity as indicated by the ribcage at the surface, is in fact still far larger than the volume contained inside. This is primarily because of the marked cranial convexity of the diaphragm, the apex of its dome projecting for some considerable way into the thorax, in the midline as far forwards as intercostal space 6 as drawing **B** indicates. *The actual volume and extent of the thoracic cavity and therefore of the lungs is considerably less than would appear to be the case on purely surface inspection.*

The outer limit of the left **pleural cavity** is shown in **B** by the broken line. Since the lining layer of parietal pleura is applied to the wall of the thoracic cavity, in surface projection thoracic and pleural cavities should coincide to a considerable extent. However, cranially the drawing shows the pleural cavity as projecting forwards in advance of the first rib through the thoracic inlet into the base of the neck as a **pleural pocket**. If you recall the musculature in this region you will remember that the scalene muscles pass from neck vertebrae back onto the first few ribs in the medial wall of the axilla (armpit). The pleural pocket lies beneath these muscles and can be approached from in front through the jugular fossa at the base of the neck. It is therefore potentially vulnerable to penetrating wounds. Also the drawing shows that caudally, where diaphragmatic musculature attaches peripherally to the internal face of the costal arch, the pleural cavity terminates some small distance in advance of the costal arch. Here costal pleura on the thoracic wall meets diaphragmatic pleura on the surface of the diaphragm at the **costodiaphragmatic line of pleural reflection**. Although admitting of some variation in position it is a cranially concave line more or less paralleling the costal arch. It runs up and back from the sternum, along the lower end of costal cartilage 8 to its bend, across the middle of costal cartilage 9, just above the costochondral junction of rib 11, and ends dorsal to the middle of the last rib. The costodiaphragmatic line of pleural reflection is of especial importance from two points of view: firstly it represents the absolute limit for caudal expansion of the lungs; secondly it is the dividing line at the surface of the body between thorax and abdomen.

What has been said so far about lungs and pleural cavities might have led you to the conclusion that lung size coincides with pleural cavity size since earlier I suggested that parietal and visceral pleura are in contact. But as you may determine from **C** at bottom left of the page the outline of the left **lung** does not correspond to the limits of the pleural cavity, although closely paralleling it. *The pleural cavity is always more extensive than the lung because there are parts of the pleural cavity where parietal* *pleural layers are in apposition occluding the cavity.* These are principally: in the pleural pocket; lateral and ventral to the heart where costal and mediastinal pleura meet across the **costomediastinal recess**; and in advance of the costodiaphragmatic line of pleural reflection where costal and diaphragmatic pleura are in contact across the **costodiaphragmatic recess**. The extent of apposition and so obliteration of these recesses depends upon the phase of respiration.

Lung size and relationships depend upon respiratory movements being made by the thorax. In heavy stippling **C** shows the surface projection of the lung after expiration (breathing out). Its apex lies medial to the first rib barely entering the pleural pocket; the thin irregular ventral border does not extend far into the costomediastinal recess and is some way from the sternum; and the basal border of the lung does not extend far back into the costodiaphragmatic recess, it follows a curved line caudodorsally from the costochondral joint of rib 6 to the vertebral end of the penultimate intercostal space. Overall lung expansion is limited by lung structure itself – the collagen component of connective tissue prevents overdistension and consequent elastic tissue disruption, but also by the absolute limits of the pleural cavity. The bulk of lung expansion will occur in a lateral direction, with outward movement of the ribcage, and a longitudinal direction, with flattening of the dome of the diaphragm. In terms of lateral surface projection the difference in outline of a lung between maximal expiration and maximal inspiration is indicated on **C**. There is a slight expansion of the lung apex into the pleural pocket, some expansion of the ventral border into the costomediastinal recess, and some caudal expansion of the basal border into the costodiaphragmatic recess. The basal border of the lung actually moves through a distance of several centimeters in full respiratory movement as the lung extends back into the costodiaphragmatic recess. But even at maximal lung expansion the recess is never fully opened up and the basal border of a normal healthy lung never reaches the costodiaphragmatic line of pleural reflection.

The outline of the lung might deviate from these parameters for a number of reasons. Its extent might be decreased by some measure of lung 'collapse' with either air (pneumothorax) or fluid (hydrothorax) in the pleural cavity, both possibilities in certain pathological, traumatic conditions. Its extent might be increased by some measure of pathological lung distension such as emphysema when elastic tissue in lung is damaged and lung tissue has become overstretched. This means that inspiration commences into a lung already partially distended, consequently scope for overall inspiration is decreased and powerful inspiratory movements occur.

In a normal standing position the **heart** lies medial to the bulk of the upper arm (**A**). It extends longitudinally between ribs 3 and 6 ('overlapping' into intercostal spaces 2 and 6). The dorsal extent of the heart is approximately on a line connecting the acromion process of the scapula (immediately lateral to the shoulder joint) with the ventral end of the last rib. Consequently the main body of the heart lies in the ventral half of the chest (**C**). It is an obliquely oriented, blunt cone-shape with a long axis at 45° to the vertical, its apex lying slightly to the left of the midline close to the sternum and related to costal cartilages 5 and 6 and the second to last sternebra.

Since the chest is deeper than it is wide and the cavity inside is considerably narrower ventrally, the heart is in close association with the chest wall ventrolaterally across the costomediastinal recess. On either side the heart is flanked by the lungs which it indents markedly so that much of the intervening layer of lung tissue between heart and chest wall is really very thin and is cut into in places as *cardiac notches* through which the heart actually contacts the chest wall. The area of chest contact is in the lower ends of intercostal spaces 3, 4 and 5 and is somewhat larger on the right side (see fig 24). On this right side the right ventricle contacts the chest wall in the notch between the cranial and middle lobes of the lung and is accessible through intercostal spaces 4 or 5 close to the sternum. You may take the beat of the heart from its apex at the lower end of intercostal space 6 or 7.

In **D** at centre right of the page the outline of the **abdominal** and **pelvic cavities** is indicated by the continuous line since these are coextensive through the pelvic inlet. The abdominal roof is formed from lumbar vertebrae and sublumbar muscles; its walls and floor from muscles of the flanks and belly; its cranial boundary is the diaphragm; and its caudal boundary is the pelvic inlet (the aperture formed from the sacrum dorsally and the cranial border of the pelvic girdle bilaterally and ventrally). The boundaries of the pelvic cavity are a little less clear: its roof is formed from the sacrum, first 3 caudal vertebrae and the ventral sacrocaudal muscles; its walls and floor from pelvic bones, sacrotuberous ligaments, internal obturator muscles and rump muscles; its cranial boundary is the pelvic inlet; its caudal boundary is the pelvic outlet (the somewhat triangular 'hole' bordered by caudal vertebra 3, the sacrotuberous ligaments and the ischiatic arch). The outlet is less confined than the inlet and can be enlarged further through lifting the tail behind the short sacrum. Internally the musculoskeletal boundaries are lined by fascia which completely encloses the cavities.

The **abdominal cavity** is roughly circular in cross section and longer than it is deep. As you may see from the drawing the cavity does not coincide with the outer limits of the abdomen: the upper limit is always some distance from the upper contour of the loins, and the considerable cranial (intrathoracic) extent (outlined by the dome of the diaphragm extending as far forwards as a plane through rib 7) should be noted since this is considerably in advance of the costal arch which marks the cranial boundary at the surface. *The apparently small surface extent of the abdomen in many dogs is therefore misleading.* The intrathoracic part of the abdominal cavity is necessarily more restricted in its capacity to change in shape and size than the more caudal part of the abdomen with its predominantly muscular walls.

The **pelvic cavity** is also roughly

circular in section as is the pelvic inlet. However, since the ilia of the pelvic girdle slope strongly down and back from their sacral articulations, the pubic brim lies ventral to the sacrocaudal boundary and the inlet is set at an oblique angle to the long axis of the body. Since the pelvic roof is longer than the floor the pelvic outlet although still oblique is more upright than the inlet. However, the longitudinal axis of the pelvic canal remains more or less horizontal.

As you can see from **D**, compared with the abdomen *the pelvic cavity is small, and considerably smaller than the surface contour of the pelvic region would suggest*. The cavity is reduced further in size by the pelvic diaphragm whose muscles close off the caudolateral part of the pelvic cavity (where the walls are lacking in musculoskeletal support) from the perineum and ischiorectal fossae caudally and laterally.

The outer contour of the **peritoneal cavity** (disposition of parietal peritoneum) is shown in **D** by the broken line. Abdominal and peritoneal cavities are coincident in surface projection but in the pelvis the peritoneal cavity is considerably less in size than the pelvic cavity since much of the caudal and lateral parts of the pelvic cavity are 'excluded' by the pelvic diaphragm. Even within the restricted pelvic cavity the terminal parts of the digestive and urogenital tracts are still not peritoneal structures but are retroperitoneal in position, embedded in loose connective tissue. The extent of this retroperitoneal component is indicated by the close stippling in **D**. The peritoneal cavity

actually terminates caudally as a series of blind-ending pockets between pelvic viscera particularly clearly illustrated in the pelvis diagrams (figs 29C & 29F).

The parietal peritoneum of the abdomen is also extended on either side caudally as the **vaginal processes** projecting through the inguinal canals into the scrotum. The cavities of the processes remain in communication with the peritoneal cavity in the abdomen throughout life and represent persistent 'flaws' in the abdominal wall always presenting the potential for herniation of abdominal contents. An additional site at which herniation is not uncommon is the umbilicus (navel).

Surface features and topographical regions
1 Scapular region (shoulder). **2** Position of shoulder joint. **3** Brachial region (arm). **4** Olecranon region. **5** Tricipital margin of arm (based on long head of triceps muscle). **6** Interscapular region (withers). **7** Thoracic vertebral region (back). **8** Lumbar region (loins). **9** Lateral boundary of iliocostal muscle (ventral extent of epaxial musculature). **10** Gluteal region (rump). **11** Coxal tuberosity region (point of haunch). **12** Hip joint region. **13** Clunial region (including ischiorectal fossa). **14** Root of tail. **15** Femoral region (thigh). **16** Cranial margin of thigh (based on sartorius muscle). **17** Presternal region (breast). **18** Sternal region (brisket). **19** Costal region (thorax, chest or rib region). **20** Cardiac region. **21** Axilla (armpit between muscles of arm and shoulder and muscles of chest wall). **22** Hypochondriac region. **23** Xiphoid region. **24** Lateral abdominal region (flank). **25** Fold of flank. **26** Umbilical region (belly). **27** Inguinal region. **28** Pubic region. **29** Preputial region (prepuce [sheath] surrounding glans penis). **30** Anus. **31** Penile (urethral) bulb (in

root of penis). **32** Scrotum.

Bones, joints and ligaments
33 Median sacral crest. **34** 1st thoracic spinous process. **35** 1st lumbar spinous process. **36** Lumbar transverse processes. **37** Sternum (breastbone). **38** Manubrium of sternum (1st sternebra elongated into base of neck). **39** Rib 1 (bordering thoracic inlet). **40** Rib 3. **41** Rib 6. **42** Rib 13 (last or floating rib). **43** Costal arch (fused costal cartilages of ribs 10–12 attached to costal cartilage of rib 9). **44** Dorsal (vertebral) border of scapula. **45** Cranial angle of scapula. **46** Caudal angle of scapula. **47** Spine of scapula. **48** Acromion process of scapula. **49** Greater tuberosity of humerus (point of shoulder). **50** Olecranon process of ulna (point of elbow). **51** Crest of ilium. **52** Coxal tuberosity of ilium (point of haunch). **53** Sacral tuberosity of ilium (point of croup). **54** Pubic pecten. **55** Ischiatic tuberosity (point of buttock). **56** Greater trochanter of femur. **57** Sacrotuberous ligament.

Surface projections of body cavities
58 Surface projection of outer boundary of thoracic cavity – endothoracic fascia lining thoracic cavity (unbroken line incorporating 1st rib [39], sternum [37], costal arch [43] and last rib [42]). **59** Contour of diaphragm in median plane of body. **60** Surface projection of outer boundary of abdominal cavity – transversalis fascia lining abdominal cavity (unbroken line incorporating diaphragm [59]). **61** Surface projection of outer boundary of pelvic cavity – pelvic fascia lining pelvic cavity (unbroken line continuous with outer abdominal boundary at pelvic inlet). **62** Thoracic inlet (bordered by 1st ribs, thoracic vertebra 1 and sternal manubrium). **63** Pelvic inlet (more or less rounded opening at an oblique angle to long axis of body and bordered above by sacral promontory, below by cranial borders

of pubic bones and on either side by arcuate lines on ilia – boundary indicated by dotted line). **64** Pelvic outlet (roughly triangular opening bordered above by caudal vertebra 3, below by ischiatic arch and ischiatic tuberosities and on either side by sacrotuberous ligaments).

Surface projection of coelomic cavities
65 Surface projection of outer boundary of left pleural cavity (parietal [costal] pleura – indicated by broken line). **66** Pleural pocket of left pleural cavity. **67** Costodiaphragmatic line of pleural reflection. **68** Position of costodiaphragmatic recess of left pleural cavity. **69** Costomediastinal line of pleural reflection. **70** Position of costomediastinal recess of left pleural cavity. **71** Surface projection of outer boundary of peritoneal cavity in abdomen and pelvis (parietal peritoneum – indicated by broken line). **72** Retroperitoneal component of pelvic cavity (indicated by dense stipple). **73** Vaginal process from peritoneal cavity in abdomen (projecting through inguinal canal).

Surface projections of thoracic viscera
74 Surface projection of left lung at maximal expiration. **75** Surface projection of left lung at maximal inspiration. **76** Position of tracheal bifurcation. **77** Heart.

Surface projections of abdominal and pelvic viscera
78 Stomach. **79** Greater curvature of stomach. **80** Fundus of stomach. **81** Transverse colon. **82** Descending colon. **83** Rectum (continuing colon through pelvic cavity). **84** Anal canal. **85** Liver. **86** Spleen. **87** Left kidney. **88** Left ureter. **89** Urinary bladder. **90** Pelvic urethra. **91** Prostate gland. **92** Root of penis. **93** Body of penis. **94–95** Glans of penis. **94** Bulbus glandis. **95** Pars longa glandis. **96** Left testis. **97** Left vas deferens.

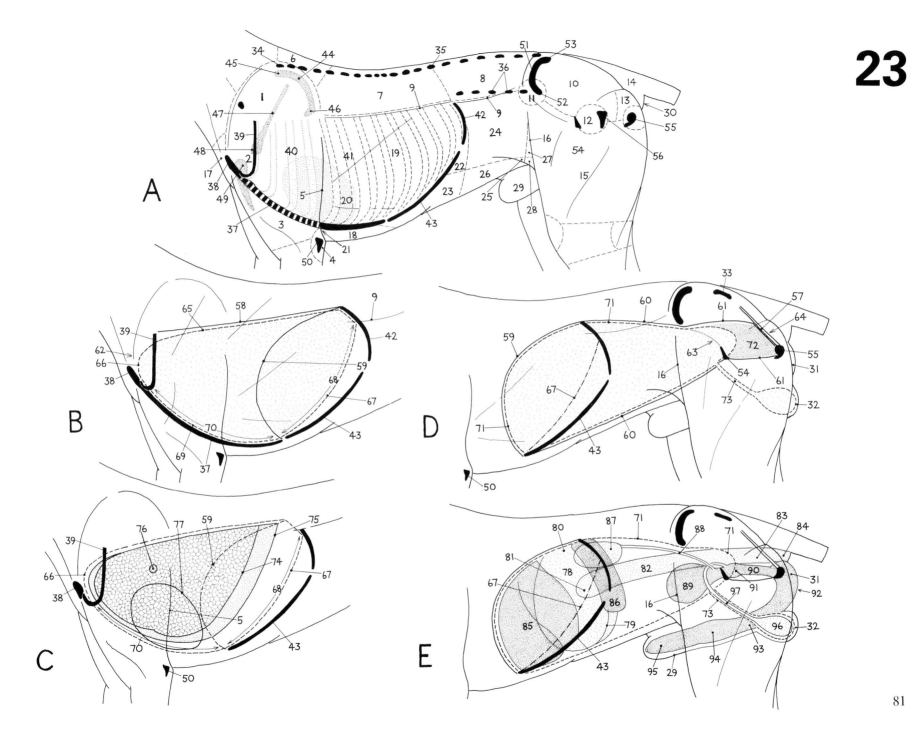

24

SURFACE PROJECTIONS OF THORACIC, ABDOMINAL AND PELVIC VISCERA OF THE BITCH FROM THE RIGHT SIDE

As a complement to the five preceding drawings of a dog from the left side, these five outline sketches are of a bitch from the right side. The significant difference in the chest is that on the right (**D**) the ventral border of the lung is indented with a **cardiac notch** through which the heart gains greater contact with the chest wall at the lower ends of intercostal spaces 4 or 5. The cavity of the heart in this area is the right ventricle.

In **E** the lung is again considered but from the point of view of auscultation — listening to the sounds emanating from a lung. You might think that lung sounds could be heard from any point at which a lung is in contact with the chest wall. Not so, however, since for instance caudally, lung tissue in the region of the basal border of the lung has little depth where it occupies the costodiaphragmatic recess. Dorsally also, the upper limit of a lung is some distance in from the body surface covered by the epaxial muscles (iliocostal and longissimus). Consequently there is a roughly triangular minimum area of constant contact between lung and chest wall mapped by three construction lines:

Cranial boundary — the tricipital margin of the upper arm extending more or less vertically from point of elbow to caudal angle of scapula, roughly cor-responding to rib 5.

Dorsal boundary — the ventral limit of the epaxial muscle mass from rib 5 caudally to intercostal space 11, corresponding to a line from the caudal angle of the scapula to the coxal tuberosity of the pelvic bone.

Caudoventral boundary — an oblique line from the upper end of the penultimate intercostal space (at the ventral border of the epaxial musculature) to the point of the elbow at the lower end of intercostal space 6, crossing the middle of rib 8.

Surface features and topographical regions
1 Scapular region (shoulder). **2** Shoulder joint region. **3** Brachial region (arm). **4** Olecranon region. **5** Tricipital margin of arm (based on long head of triceps muscle and in a normal standing position approximately at the level of rib 5). **6** Interscapular region (withers). **7** Thoracic vertebral region (back). **8** Lumbar region (loins). **9** Lateral boundary of iliocostal muscle (ventral extent of epaxial musculature). **10** Gluteal region (rump). **11** Coxal tuberosity region (haunch). **12** Hip joint region. **13** Clunial region (including ischiorectal fossa). **14** Root of tail. **15** Femoral region (thigh). **16** Cranial margin of thigh (based on sartorius muscle). **17** Pre-sternal region (breast). **18** Sternal region (brisket). **19** Costal region (thorax, chest or rib region). **20** Cardiac region. **21** Axilla (armpit between muscles of arm and shoulder and muscles of chest wall). **22** Hypochondriac region. **23** Xiphoid region. **24** Lateral abdominal region (flank). **25** Fold of flank. **26** Umbilical region. **27** Inguinal region. **28** Pubic region. **29** Anus.

Bones, joints and ligaments
30 Median sacral crest. **31** 1st thoracic spinous process. **32** 1st lumbar spinous process. **33** Lumbar transverse processes. **34** Sternum (breastbone). **35** Manubrium of sternum (sternebra 1 elongated into base of neck). **36** Rib 1 bordering thoracic inlet. **37** Rib 13 (last or floating rib). **38** Costal arch (fused costal cartilages of ribs 10–12 attached to costal cartilage of rib 9). **39** Dorsal (vertebral) border of scapula. **40** Cranial angle of scapula. **41** Caudal angle of scapula. **42** Spine of scapula. **43** Acromion process of scapula. **44** Greater tuberosity of humerus (point of shoulder). **45** Olecranon process of ulna (point of elbow). **46** Crest of ilium. **47** Coxal tuberosity of ilium (point of haunch). **48** Sacral tuberosity of ilium (point of croup). **49** Pubic pecten. **50** Ischiatic tuberosity (point of buttock). **51** Greater trochanter of femur.

Surface projections of body cavities
52 Surface projection of outer boundary of thoracic cavity — endothoracic fascia lining thoracic cavity (unbroken line incorporating 1st rib [36], sternum [34], costal arch [38] and last rib [37]). **53** Contour of diaphragm in median plane of body. **54** Surface projection of outer boundary of abdominal and pelvic cavities — transversalis and pelvic fascia lining abdominal and pelvic cavities (unbroken line incorporating diaphragm [53]). **55** Thoracic inlet (bordered by 1st ribs, thoracic vertebra 1 and sternal manubrium). **56** Pelvic inlet (more or less rounded opening bordered above by sacral promontory, below by cranial borders of pubic bones and on either side by arcuate lines on ilia). **57** Pelvic outlet (roughly triangular opening bordered above by caudal vertebra 3, below by ischiatic arch and ischiatic tuberosities and on either side by sacrotuberous ligaments).

Surface projections of coelomic cavities
58 Surface projection of outer boundary of right pleural cavity (parietal [costal] pleura – broken line). **59** Pleural pocket of right pleural cavity. **60** Costodiaphragmatic line of pleural reflection. **61** Position of costodiaphragmatic recess of right pleural cavity. **62** Costomediastinal line of pleural reflection. **63** Position of costomediastinal recess of right pleural cavity. **64** Surface projection of outer boundary of peritoneal cavity in abdomen and pelvis (parietal peritoneum – broken line). **65** Retroperitoneal component of pelvic cavity (dense stipple).

Surface projections of thoracic viscera
66 Surface projection of right lung at maximal expiration (dense stipple). **67** Area of auscultation of lung. **68**–**70** Borders of auscultatory area. **68** Cranial border (tricipital margin in standing animal, from caudal angle of scapula to point of elbow roughly corresponding to rib 5). **69** Dorsal border (lower border of epaxial musculature from rib 5 caudally to intercostal space 11 approximately on a line from caudal angle of scapula to coxal tuberosity roughly paralleling vertebral column). **70** Caudoventral (basal) border (from point of elbow at lower end of intercostal space 6 to upper end of intercostal space 11 and crossing middle of rib 8). **71** Heart. **72** Cardiac notch in ventral border of right lung.

Surface projections of abdominal and pelvic viscera
73 Duodenal loop. **74** Ascending duodenum. **75** Duodenojejunal flexure. **76** Liver. **77** Right kidney. **78** Right ureter. **79** Urinary bladder. **80** Neck of bladder. **81** Urethra. **82** Right ovary. **83** Right uterine horn. **84** Uterine body. **85** Position of uterine cervix. **86** Vagina. **87** Vestibule. **88** Vulva.

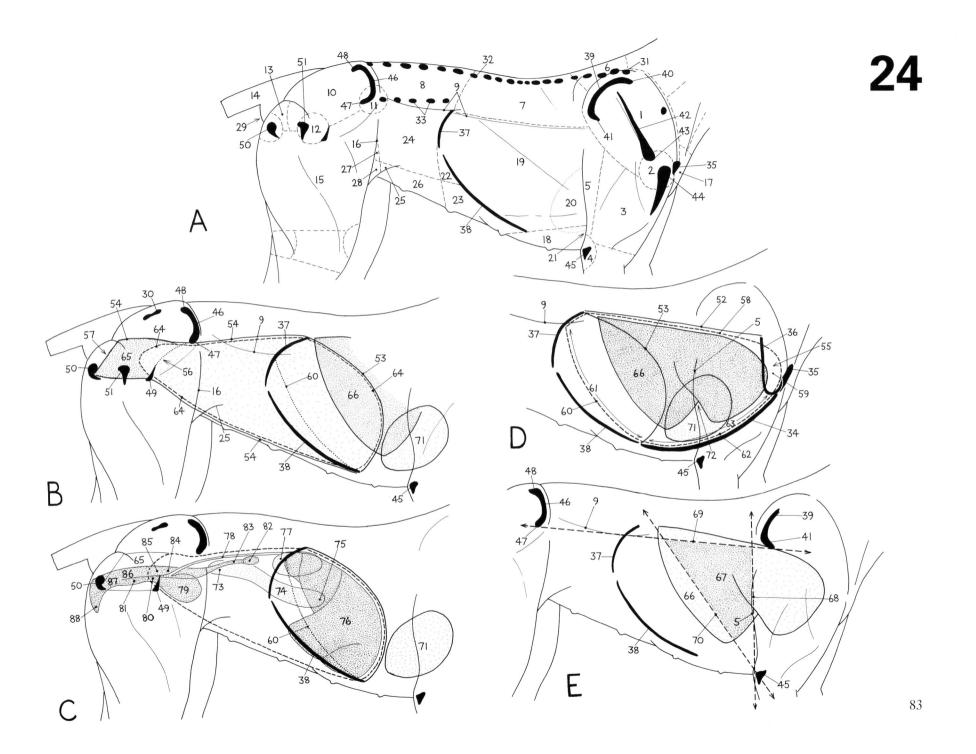

25

HEAD OF THE DOG – SKULL, SURFACE FEATURES OF THE MOUTH AND TEETH AND SURFACE PROJECTION OF THE SALIVARY GLANDS

The head is shown in some detail in a number of the drawings. The selection here is concerned predominantly with digestive structures located in the head – lips, mouth, teeth, tongue and salivary glands.

The **lips** surround the entrance into the mouth cavity and are fairly lengthy structures so that the commissures at the angles of the mouth are some way back on either side. Consequently the **cheeks** which form the remainder of the outer boundary of the **oral vestibule** (slit-like space outside of the teeth and gums) are small. Long lips and restricted cheeks means that the gape of the mouth can be quite considerable (**D**). The muscular basis of both lips and cheeks is principally the buccinator and orbicularis oris of the facial muscles. As well as keeping food and drink in the mouth these also enable the lips to change shape and move to impart 'expression' to the face. The **oral cavity proper** is internal to the teeth and houses the tongue. When the jaws are closed it communicates with the vestibule through interdental spaces between teeth and retromolar spaces caudal to the last cheek teeth.

The shape, number and position of the teeth are shown in several of the drawings, and I'm sure that you are familiar with many of the details of your dog's teeth.

Incisor teeth – 'simple', small chisel-like teeth rather loosely embedded in the jaws rostrally. 6 incisors in both upper and lower jaws are closely grouped together and are useful in cutting, nipping, grasping and grooming.

Canine (eye) teeth – large, curved, laterally compressed but simple in shape and deeply rooted. Useful as aggressive weapons, stabbing and ripping, and for holding.

Premolar and molar teeth – the more complex cheek teeth (the distinction between the two types based on the premolars having milk precursors). 8 premolars in both upper and lower jaws increasing in size and complexity from first to last. 4 molars in the upper jaw and 6 in the lower, decreasing in size from first to last. Useful in shearing and grinding.

As a general rule dogs do not chew their food to any great extent ('gnawing' on a bone is not quite the same thing), they tend to bolt it down rapidly. It is not necessary therefore for upper and lower teeth to meet (occlude) to any great extent and flattened grinding surfaces are only present on the rear molar teeth. Most of the cheek teeth are flattened from side to side and have sharp longitudinally aligned cusps producing cutting edges. At the same time the upper jaw and upper dental arch are slightly wider than the lower allowing certain of the upper teeth to 'bite' against the outer surfaces of the lower teeth in a shearing scissor-like action.

In a 'normal' dentition, in a dog of mesaticephalic type, only the rostral (incisors) and the caudal cheek teeth (rear molars) actually meet when the jaws are shut. The massive canines overlap and are used to grab and hold onto objects, while the carnassial or sectorial teeth (upper premolar 4 and lower molar 1) in the cheek also overlap and are used as shears to cut and tear through tendon and bone. Between the canines and carnassials upper and lower teeth do not meet at all the tooth crowns of each row in this space are aligned longitudinally to form a continuous serrated cutting edge. These teeth are used for grasping, puncturing and tearing and for carrying things (the space often being called the 'premolar carrying space'). Apart from the first lower molar tooth the remaining molar crowns are irregularly flattened and used for the limited amount of crushing and grinding that a dog can manage. The two upper molar teeth are also orientated transversely rather than rostrocaudally.

The sketch in the centre of the page (**E**) shows a skull from which the bone forming the outer walls of the tooth sockets has been removed to display the extent of the tooth roots. Incisor and canine teeth, the first upper and lower premolar teeth and the last lower molar tooth are all single rooted – the enormous root of the upper canine tooth can be felt where it is embedded in the jaw bone. Practically all of the remaining cheek teeth are double rooted bar upper premolar 4 and upper molars 1 and 2, all of which have an additional third root.

Finally, when considering teeth we must not forget that dogs have two sets of teeth:

A temporary set of small 'milk' teeth suitable in size and strength for the jaws of a puppy.

A larger and stronger permanent set of teeth in a grown dog.

Milk teeth appear from three weeks onwards resembling permanent teeth but smaller and sharper and with long slender roots. Permanent teeth are stronger and better adapted to the larger jaws and more vigorous action of the grown dog. All milk teeth are replaced as the jaws grow, and extra new teeth are added between the canine and permanent premolar teeth (premolar 1 erupts several weeks later than the deciduous teeth and is considered as part of the permanent dentition), and at the rear end of the jaws, the molar teeth. The temporary set are fully functional by six to eight weeks, replacement by the permanent set beginning at nine to ten weeks and being completed by six to seven months.

The sketch at bottom right of the page (**F**) shows certain of the glandular structures in the head in surface projection. The **major salivary glands** are the *parotid, mandibular, sublingual and zygomatic*, producing a continuous flow of saliva which increases when food is introduced into the mouth. Saliva has no digestive role *per se* but it cleans and moistens the mouth and assists in swallowing by lubricating boluses of food. It also dissolves water soluble components of food facilitating their access to the taste buds on the tongue. Saliva does contain some excreted compounds which may accumulate as tartar on the teeth. A number of small clusters of primarily mucous secreting units are freely distributed beneath the mucosa of the lips, cheeks, tongue and palate. Individually these **minor salivary glands** are unimportant (bar the *ventral buccals* in the cheek) but their collective contribution is considerable.

The ducts of the parotid and zygomatic salivary glands open into the oral vestibule in the cheek: the parotid on a raised papilla opposite upper premolar tooth 4, the zygomatic a little further caudally opposite the first upper molar tooth. However, the ducts of the mandibular and the monostomatic component of the sublingual gland open on a sublingual papilla in the oral cavity proper beneath the tongue behind the lower incisors. Raise the tongue and you will see these papillae on either side of the frenulum of the tongue the mucosal fold joining tongue with floor of mouth. Distinct sublingual folds of mucous membrane run back in the floor of the mouth from these papillae. Food taken into the mouth is 'manipulated' by the tongue and teeth and moistened by saliva before it is swallowed. The **tongue** is a large, highly muscular and therefore very mobile organ supported in the floor of the mouth and pharynx by a series of small bones, the hyoid apparatus. With the mouth closed the tongue practically fills the oral cavity (fig 26D).

The muzzle contains the movable part of the nose based on **nasal cartilages** attached to the margins of the bony nasal aperture. These surround the comma-shaped **external nostrils** maintaining their patency. A very limited amount of nostril dilation is brought about by those facial muscles which attach onto the nasal cartilages. The curvature of the nostril is produced by a bulbous dorsolateral *alar fold* – a continuation of the mucous membrane covering the rostral end of the ventral nasal concha that fuses with the inner side of the nostril wing largely filling the **nasal vestibule** within the confines of the nasal cartilages (see fig 26D). This fold diverts inspired air medially and ventrally, and in some brachycephalic dogs is large enough to hinder breathing. When a dog breathes in respiratory air flow enters its nasal cavity, passes along channels (**meatuses**) between the nasal conchae, leaves through **internal nostrils** above the rear end of the hard palate and enters the nasopharynx above the soft palate (see figs 26C & 26E).

The skin surrounding the nostrils is hairless and heavily pigmented as the so-called **nasal plane**. It is tough and moist with a pattern of ridges giving a 'nose-print' analogous to our own fingerprints and probably just as characteristic. Since this skin is non-glandular its moisture in a healthy dog is largely due to overflow from numerous serous and mucous glands of the nasal vestibule and nasal cavity generally. However, a significant contribution comes from **lacrimal** and **lateral nasal glands** whose ducts open into the vestibule: the nasolacrimal duct in the floor through a tiny hole ventral to the alar fold, the duct of the lateral nasal gland dorsocaudal to the alar fold. The position of these glands and ducts is shown in F.

Since the **jaws** and teeth are the prime determinants of facial shape the nose and the entire nasal apparatus, although major components of the face, must fit in with the structural framework determined by the jaws. A factor of some considerable importance when selective breeding reduces facial length. Also influencing the appearance of the face are the **eyes**. The eyeballs being mobile and vulnerable structures are housed in sockets (orbits) protected by a bony rim (**C**) completed caudally by an **orbital ligament**. Orbital openings face outwards and forwards and are more rounded in short-headed breeds. The transparent corneal surface of the eyeball (which allows light to enter the interior of the eye) projects beyond the bony protective rim and is shielded by **eyelids**. The lids meet at inner and outer angles and therefore surround a palpebral fissure (**B**). Both eyelids have an outer covering of normal, hairy skin, whilst the upper also supports a row of protective eyelashes (cilia). Some support is given to the eyelids by an internal thickening of dense fibrous tissue, deep to which sebaceous **tarsal glands** open onto the edge of the lid through a row of tiny holes along with ciliary glands (modified sweat glands).

Internally the lids are covered with a thin, delicate moist membrane, the **conjunctiva**, continuous with a similar conjunctival layer covering the exposed corneal surface of the eyeball. The conjunctival sac is the potential space bounded by these two layers into which lacrimal fluid (tears) is shed from the gland at the upper outer margin of the eye below the supraorbital process of the frontal bone (**F**). The watery secretion flows down across the eyeball keeping conjunctival membranes moist and cleaning the corneal surface. Eventually tears drain away through small openings in the edges of the eyelids close to the eye's medial angle. These openings lead into ducts which transport the fluid into the **nasolacrimal duct**, a tube penetrating the lacrimal bone in the inner wall of the orbit to pass into the nasal cavity. The duct eventually leads into the floor of the nasal vestibule just inside the nostril. To prevent normal flow of lacrimal fluid from spilling over onto the face, the tarsal glands in the eyelids produce a viscid, oily fluid.

Unlike ourselves a dog has an additional **third eyelid** (nictitating membrane) at the inner angle of its eye. This lid is covered completely by conjunctiva, in fact it is really only an extra fold of conjunctiva. It seems to be an additional mechanism for sweeping across the eyeball and cleansing it. Normal eyeball prominence, a result of outward pressure from intraorbital contents, keeps the third eyelid in a 'retracted' position. Upper and lower eyelids are moved by muscles which lie directly inside them: the orbicularis oculi sphincter of muscle closes the palpebral fissure; the levator muscle of the upper eyelid opens the fissure. The third eyelid on the other hand does not have muscles directly associated with it but is moved by a T-shaped cartilage – the cross bar is located in the free edge of the eyelid, the stem projects back into the orbit alongside the eyeball. It operates somewhat as follows: attached to the rear of the eyeball in the depths of the orbit is a four part *retractor muscle* which can pull the eyeball deeper into its socket. Such movement squeezes the fat that is normally present filling the spaces between muscles in the socket and displaces it forwards around the eyeball. The movement of fat pushes on the stem of the T which moves the third eyelid across the cornea. In a starved, emaciated dog, orbital fat has often been used up so its eyeball becomes sunken and its third eyelid is allowed to protrude permanently across its eyeball.

Surface features
1 Nasal plane. 2 External nostril (leading into nasal vestibule surrounded by nasal cartilages). 3–7 Lips (surrounding mouth opening – oral fissure). 3 Upper lip (supporting superior labial sensory hairs). 4 Lower lip. 5 Philtrum. 6 Wing of nostril. 7 Commissure of lips (at angle of mouth). 8 Prominence of chin (mentum – supporting mental sensory hairs). 9 Sensory hairs of face (supraorbital, zygomatic, buccal, intermandibular). 10 Foreface. 11 Stop. 12 Forehead. 13 Cheek (based on buccinator muscle). 14 Eyeball (situated in orbit and protected by bony orbital rim – outline of eyeball indicated by broken line in orbit in fig C). 15–17 Eyelids (surrounding palpebral fissure). 15 Upper eyelid (supporting cilia – eyelashes). 16 Lower eyelid. 17 Third eyelid (nictitating membrane). 18 Position of lacrimal puncta in eyelids. 19 Lacrimal caruncle. 20 Medial angle of eye. 21 Lateral angle of eye. 22 Iris (visible through cornea). 23 Sclera (white of eye). 24 Pinna of ear (visible part of external ear). 25 Helix (scapha of auricular cartilage). 26 Cranial margin of helix with spine. 27 Marginal cutaneous pouch of pinna. 28 Tragi (prominent hairs at opening of ear canal). 29 External opening of ear canal (facing dorsally). 30 Anthelix. 31 Pretragic notch. 32 Intertragic notch. 33 Tragus of conchal part of auricular cartilage. 34–37 Tongue. 34 Apex of tongue. 35 Body of tongue. 36 Root of tongue. 37 Circumvallate taste buds of tongue. 38 Lingual frenulum. 39 Palatoglossal fold. 40 Palatine tonsil (in tonsillar fossa). 41 Epiglottis (based on epiglottic cartilage). 42 Soft palate. 43 Palatine ridges of hard palate. 44 Sublingual papilla (opening of mandibular and sublingual salivary gland ducts). 45 Sublingual fold of mucosa (on course of salivary gland ducts). 46 Gums (gingivae). 47 Carnassial (shearing) teeth (upper premolar 4 and lower molar 1). 48 Jugular groove. 49 Neck. 50 Throat.

Bones
51 Facial region of skull (based on nasal cavity and jaws). 52 Nasal bone. 53 Nasal process of incisive bone (bordering entry into bony part of nasal cavity). 54 Maxillary bone (forming much of lateral surface of muzzle). 55 Infraorbital foramen. 56 Alveolar border of maxillary bone (bearing teeth of upper dental arch). 57–63 Mandible (lower jaw). 57 Body of mandible. 58 Angular process of mandible. 59 Masseteric fossa of mandibular ramus (insertion of masseter muscle). 60 Coronoid process of mandible (insertion of temporal muscle). 61 Alveolar border of mandible (bearing teeth of lower dental arch). 62 Mandibular symphysis. 63 Mental foramina. 64 Cranial region of skull (braincase). 65 Orbit (housing and protecting eyeball). 66 Bony orbital rim (from frontal, maxillary and zygomatic bones). 67 Zygomatic process of frontal bone. 68 Zygomatic arch (bridge of bone connecting face and cranium below eye). 69 Orbital ligament (completing orbital rim). 70 Temporal line of frontal bone (rostral divergence of external sagittal crest). 71 External sagittal crest (in dorsal midline of cranium). 72 External occipital protuberance (occiput). 73 Nuchal crest (dividing caudal from lateral surface of cranium). 74 Mastoid process of temporal bone (sole representation on skull surface of petrous temporal bone). 75 Temporomandibular joint (position shown by broken circle in fig A). 76–78 Hyoid apparatus (support for tongue and larynx in floor of throat). 76 Cranial horn of hyoid (in wall of pharynx – composed of tympanohyoid cartilage, epihyoid, stylohyoid and ceratohyoid bones). 77 Basihyoid bone (transverse bar of bone located in musculature of tongue root and linking horns of left and right sides). 78 Thyrohyoid bone (caudal horn of hyoid in wall of laryngopharynx linking basihyoid with thyroid cartilage of larynx). 79–81 Laryngeal cartilages. 79 Epiglottic cartilage. 80 Thyroid cartilage (most prominent with ventral border forming laryngeal prominence of 'voice box'). 81 Cricoid cartilage. 83 Dorsal arch of atlas. 84 Lateral vertebral foramen of atlas (intervertebral foramen 1 for passage of 1st cervical spinal nerve and vertebral artery). 85 Wing of atlas. 86 Spinous process of axis. 87 Transverse foramina of transverse canal (passage of vertebral blood vessels). 88 Atlantooccipital joint. 89 Atlantoaxial joint.

Nasal and auricular cartilages
90–93 Nasal cartilages (surrounding vestibule and supporting mobile part of nose). 90 Nasal septum. 91 Dorsolateral nasal cartilage. 92 Ventrolateral nasal cartilage. 93 Accessory nasal cartilage. 94–98 Auricular cartilage (surrounding ear canal and supporting pinna of ear). 94 Scapha of auricular cartilage (outer leaf-like portion). 95 Antitragus of caudal border of helix. 96 Medial crus of helix. 97 Lateral crus of helix. 98 Concha of auricular cartilage (basal rolled tubular portion). 99 Position of external acoustic meatus (at base of concha facing laterally across which eardrum stretched in life – shown in broken line in fig C). 100 Scutiform cartilage (lying on temporal muscle rostral to ear).

Muscles
101–102 Jaw closure muscles. 101 Temporal muscle. 102 Masseter muscle.

Glandular structures of head in surface projection
103 Parotid salivary gland (surrounding concha of auricular cartilage). 104 Parotid salivary gland duct (opening into oral vestibule through cheek opposite upper premolar 4). 105 Mandibular salivary gland. 106 Mandibular salivary gland duct (opening at rostral end of floor of oral cavity proper on sublingual caruncle to side of lingual frenulum). 107 Sublingual salivary gland, monostomatic part. 108 Sublingual salivary gland duct (accompanies duct of mandibular gland). 109 Sublingual gland, polystomatic part (diffuse lobules alongside mandibular duct opening separately in floor of mouth). 110–111 Buccal salivary glands. 110 Ventral buccal glands (intermingled with fibres of buccinator and orbicularis oris muscles in cheek). 111 Zygomatic gland (consolidation of dorsal buccal glands in base of orbit below eyeball emptying into oral vestibule caudal to parotid duct). 112 Lacrimal gland (beneath supraorbital process of frontal bone and emptying into conjunctival sac). 113 Nasolacrimal duct (leading from lacrimal ducts from lacrimal puncta at medial angle of eye – runs through wall of nasal cavity to empty into nasal vestibule below alar fold). 114 Lateral nasal gland (located in wall of maxillary recess). 115 Duct of lateral nasal gland (opening in nasal vestibule above alar fold).

Teeth
I1–I3 Incisor teeth 1–3. C Canine tooth. P1–P4 Premolar teeth 1–4. M1–M3 Molar teeth (1–2 upper and 1–3 lower).
(D after Taylor, 1955)

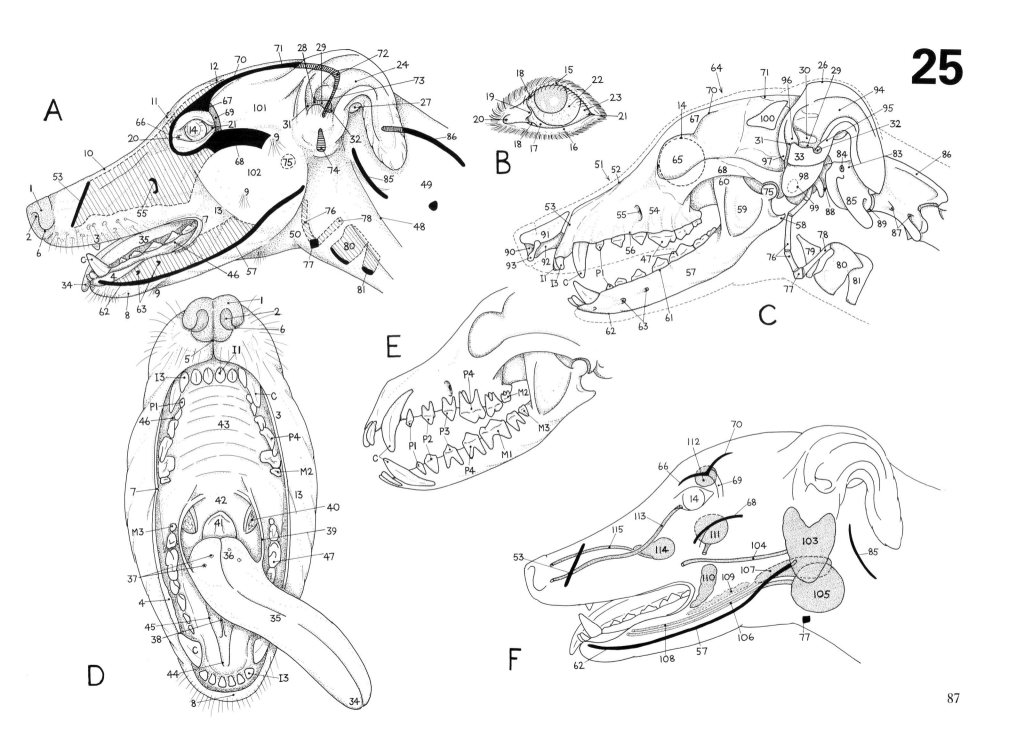

25

87

26

DEEP STRUCTURES OF THE HEAD OF THE DOG AND SURFACE PROJECTIONS OF THE CRANIAL, NASAL, ORAL AND PHARYNGEAL CAVITIES

The drawings on this page are concerned with structures deep inside the head and so in the main must be visualized by you rather than you being able to feel them directly. At bottom left of the page **D** shows a midline section of the whole head and cranial end of the neck viewed from the inside. The cavities displayed in the skull drawing (fig 5B) are now occupied by soft tissues of various sorts. The remaining drawings on the page are surface views of the head onto which have been superimposed the outlines of these cavities. **A**, **B** and **C** show the nasal cavity and paranasal sinuses, and **C** also shows the cavity of the middle ear with its connection to the nasopharynx: **E** shows the cranial, oral, pharyngeal and laryngeal cavities.

The bones enclosing the **nasal cavity** form much of the bony surface of the face in front of the eyes and much of this bone is palpable from the surface. Inside the nasal cavity a sensory epithelium containing receptor organs associated with the sense of smell is supported on a complex ethmoid bone. Since this epithelium (and in particular the smell receptors) is fragile, air to be smelled must be at the correct tempera-

ture and humidity, and should be free of dust particles. Hence some of the remaining nasal epithelium is concerned with filtering, moistening and heating/cooling air that is to be smelled. This 'air-conditioning' epithelium is found in the more rostral parts of the cavity, the olfactory epithelium itself being confined to its caudal part, and is supported on **nasal conchae**, flanges of bone which have grown inwards from the nasal and maxillary bones of the cavity wall.

The nasal cavity is large, and in fact longer than would appear from the surface. Rostrally the cavity is more or less tubular, caudally it widens and deepens although both rostral and caudal openings are constricted. Since its caudal boundary is curved from side to side it extends further caudally on either side than in the midline (to a level through the medial angle of the eye). You can visualize from this description that the nasal cavity extends back for some distance between the orbits. Caudoventrally the nasal cavity extends back to the caudal border of the hard palate: caudodorsally it extends to the nasofrontal suture (stop) level with the middle of the orbit. In width the cavity is somewhat narrower than would appear from the surface since much of its caudolateral part is occupied by a maxillary recess.

Paranasal sinuses are two pairs of air-filled cavities opening into the nasal cavities. *Frontal sinuses* lie within the frontal bones; *maxillary recesses* within the maxillary bones. In surface projection (**A**) a frontal sinus is roughly triangular with its base in the dorsal midline and its apex approximately at

the zygomatic process of the frontal bone. The caudal border of the triangle follows the temporal line of the skull back to the midline of the head: the rostral border extends rostromedially to the midline just in advance of the middle of the orbit. These sinuses are normally variable in size and enlarge as a dog grows. The variation is accentuated when selective breeding alters skull length as the sketches in **B** show. The volume of the sinus is progressively reduced as the face shortens: conversely the sinus is a large and constant feature of old dolichocephalic heads where its caudal extent may reach as far back as a transverse plane through the jaw joints.

The maxillary recess occupies the face in front of the orbit above the roots of the cheek teeth. It is located medial to the infraorbital and nasolacrimal canals, both of which protrude slightly into it. Dorsally it extends to the level of the medial angle of the eye, and its opening into the nasal cavity is level with upper premolar tooth 4. The recess tends to be smaller in brachycephalic breeds as you would expect (**B**).

The **cranial cavity** corresponds in form and size closely with the brain (**D**) and depending upon breed, cranial capacity could be anywhere between 40 and 140 ccs in the adult animal. The cavity is smaller than would be thought on surface examination since paranasal sinuses, ridges for muscle attachment and temporal muscles all contribute to the head contour (**E**). Movement of the lower jaw requires large and powerful muscles whose attachments occupy large areas of the cranial surface and zygomatic arch. The major jaw muscle, the

temporal, gives its name to the temporal fossa of the skull since it occupies this whole area. In the dorsal midline, the cranial surface is normally raised into a sagittal crest to provide additional areas for temporal muscle attachment. Thus much of the external cranial architecture is directly due to jaw muscle attachment.

The remaining cavities in the head are associated with the respiratory and digestive tracts and the related skull structures of jaws, palate, hyoid apparatus and larynx. The roof of the mouth and the floor of the nasal cavity are based on the **hard palate** with the **internal nostrils** opening someway back in the mouth. To separate the nasal region from the oral region to an even greater extent, and move the nasal openings even further back in the mouth, a musculomembranous sheet is attached to the rear of the hard palate as the **soft palate**. This tissue fold is attached to the pterygoid bones on either side where they border a channel on the underside of the cranium below the orbits. The soft palate converts this channel into a tunnel, **nasopharynx**, leading back from the nasal cavity.

The shape and position of the hyoid bones are shown in relation to the skull in fig 25C and the relationship of the hyoid apparatus to surrounding throat muscles was shown in several of the muscle drawings earlier. Its long cranial elements lie in the pharyngeal walls on either side; its ventral, transverse element (basihyoid) lies in the root of the tongue in the pharyngeal floor; its shorter caudally projecting elements (thyrohyoids) also lie in the pharyngeal walls and articulate with the thyroid cartilage of the larynx. Its functions are predominantly concerned with support and suspension of the tongue in the floor of the mouth, the bulk of the tongue muscles actually originating from it, and the basihyoid lying in the tongue root. Secondarily it provides suspension and anchorage for the larynx in the floor of the throat. During swallowing the hyoid arch is swung forwards and downwards pulling the larynx along with it and so opening out the entrance into the oesophagus. The relationships between the oesophagus and larynx and the pharynx and oral cavity can be seen in the sagittal section of the head (**D**) and in the surface projection (**E**). Numerous muscle bundles inside the tongue enable it to perform the varied movements necessary for such activities as lapping fluid and manipulating food in the mouth. Several further muscles enter the tongue from attachments on the lower jaw (eg. the genioglossus), in addition to those like the hyoglossus and styloglossus arising from the hyoid apparatus, and produce those grosser movements such as protrusion and retraction which play a prominent part in lapping and swallowing.

The oral cavity proper widens from front to back like the nasal cavity but then narrows at the **palatoglossal folds**, mucous membrane folds running between tongue and soft palate on either side. This is the entry into the oral part of the pharynx and the folds are only really noticeable if the tongue is pulled out of the mouth to one side, as it is in fig 25D. Immediately caudal to the folds in the walls of the oropharynx are the prominent, elongated **palatine tonsils** located in tonsillar fossae, indentations in the pharyngeal wall.

Although quite prominent structures, normally in the grown dog tonsils are not readily apparent being hidden by an overhanging fold of mucous membrane. However, like all lymphoid tissue they are larger in infancy, in a puppy protruding quite markedly from the fossa. They are also enlarged when inflamed and excessively enlarged tonsils may impede airflow. Palatine tonsils are in fact the dominant components of a ring of lymphoid tissue in the pharynx including a pharyngeal tonsil in the dorsal midline of the nasopharynx, a lingual tonsil in the root of the tongue, and numerous small collections of lymphoid tissue in the soft palate. This ring provides the lymphoid system with opportunities for early encounters with potentially infectious agents against which antibodies can be produced.

The roof of the **oropharynx** is the soft palate, the musculomembranous fold projecting back from the hard palate, separating the nasopharynx above from the oropharynx below. Air and food tracts are therefore separated back as far as the caudal border of the soft palate where the two tracts join as the **laryngopharynx (common pharynx)**. It might be useful to consider the remaining parts of the pharynx and larynx in terms of the respiratory air flow and the food tract.

Air is drawn into the nasal cavity through the external nostrils and passes back along nasal meatuses *en route* for the lungs. At the rear of the cavity, below the ethmoid region, respiratory air flow in the nasopharyngeal meatus passes through internal nostrils into the nasopharynx below the orbit, bounded by the palatine and pterygoid bones above and the soft palate below. At the caudal boundary of the soft palate air leaves the nasopharynx and enters the laryngopharynx. Passing through the laryngopharynx, respiratory air enters the trachea (windpipe) through the *glottic opening of the larynx*. Food in the mouth cavity, having been dealt with by the teeth, is swallowed. Food is initially pushed back into the oropharynx by the tongue, lubricated in its passage by saliva. At the caudal boundary of the soft palate the food enters the laryngopharynx to reach the opening into the oesophagus.

The laryngopharynx is therefore a common passageway for both air and food and a mechanism must be present for dividing off the airway from the foodway so that *either* swallowing *or* breathing may occur, *but not both at the same time*. This separation is brought about by a cartilaginous flap, the **epiglottis**, projecting up from the pharyngeal floor and contacting the rear of the soft palate. Through this association the mouth cavity and oropharynx are effectively closed off from the laryngopharynx allowing inspired air to pass through it. Air being breathed out takes the same route back through the laryngopharynx into the nasopharynx and on through the nasal cavity. However, a dog readily resorts to mouth breathing, as we can, particularly when it is out of breath and panting. Since the epiglottis is applied to the under-surface of the soft palate there is little problem in moving it downwards out of the air flow.

When a dog is swallowing food it is pushed back in the oropharynx by the plunger-like action of the tongue root and as it enters the laryngopharynx the soft palate is raised by activity within such muscles as the palatine levator, thus opening out the oropharynx whilst at the same time closing off the nasopharynx. The hyoid apparatus, and thus the epiglottis and larynx attached to the basihyoid, is swung downwards and forwards by contraction of the geniohyoid muscles passing from the basihyoid forwards to the mandible. Downward and forward movement of the hyoid is coupled with upward and backward movement of the tongue root compressing the epiglottis over the opening into the larynx. Hyoid movement also pulls the trachea along with it thus opening out the entrance into the oesophagus. Swallowed food now has a clear passageway through the laryngopharynx into the oesophagus. But, when food and fluid are passing back through the laryngopharynx during swallowing there is always the possibility that they may pass inadvertently into the trachea through the **glottis** – 'going down the wrong way'. The entry into the trachea is therefore modified to produce a valve around the glottic opening – the **larynx** based on several movable *laryngeal cartilages*. During swallowing the glottic opening bordered by the **vocal folds** of mucous membrane must be closed by action within sphincter muscles attaching to and moving arytenoid cartilages. This system of cartilages forms the framework of the larynx (voice box) as shown in **D**.

Glottic closure is necessary for effective swallowing without inundation, but can be of use in other ways. For instance a closed glottis in conjunction with a forcible attempt to breathe out by abdominal tension is the basis of straining activities. Air cannot escape through the glottis and is retained in the trachea and lungs raising intrathoracic pressure. In turn this raises pressure in the abdomen and pelvis necessary for evacuative procedures. Sudden opening of the glottis after such a pressure rise is also the basis of coughing. Once swallowing has occurred the various pharyngeal structures return to their original positions, the glottis opens, and breathing movements can recommence.

Although the trachea can be felt through the skin in the ventral midline of the neck, it is still covered by a layer of muscles of the undersurface of the neck (sternohyoid and sternothyroid) running forwards from sternum and first costal cartilage to the hyoid apparatus and larynx in the throat just behind and between the angles of the lower jaws. They form a continuation of the strap muscles of the abdomen and thorax and in turn are continued forwards in the floor of the mouth beneath the tongue by geniohyoid muscles extending from hyoid to chin. These muscles associated with the hyoid arch and larynx will assist in moving it backwards and forwards in the throat during the swallowing movements which have been outlined above.

Bones of skull

1 Nasal cartilage (surrounding nasal vestibule). 2 Bony nasal aperture (leading into bony part of nasal cavity). 3 Nasal bone. 4 Infraorbital foramen (passage of infraorbital vessels and nerves). 5 Zygomatic arch (bar of bone linking cranium and face below eye). 6 Alveolar border of maxillary bone (upper jaw – supporting teeth of upper dental arch). 7 Hard palate. 8–10 Nasal conchae (turbinate bones – supporting nasal mucous membrane). 8 Dorsal nasal concha (nasoturbinate). 9 Ventral nasal concha (maxilloturbinate). 10 Alar fold of ventral nasal concha. 11–12 Ethmoid bone. 11 Cribriform plate of ethmoid bone (separating nasal from cranial cavities). 12 Ethmoidal labyrinth (ethmoidal conchae – attached to cribriform plate and supporting olfactory mucous membrane). 13–17 Lower jaw (mandible). 13 Body of mandible. 14 Ramus of mandible. 15 Angular process of mandible. 16 Mandibular symphysis. 17 Alveolar border of mandible (supporting teeth of lower dental arch). 18 Frontal bone. 19 Zygomatic process of frontal bone. 20 Parietal bone. 21 Occipital bone. 22 External occipital protuberance. 23 Nuchal crest. 24 Canal for transverse venous sinus (in occipital bone). 25 Cerebellar tentorium. 26 Sphenoid bone in braincase floor. 27 Pituitary (hypophyseal) fossa (in braincase floor). 28 Cranial cavity (shown in surface projection in E). 29 Foramen magnum. 30–32 Cranial horn of hyoid (situated in pharyngeal wall). 30 Epihyoid bone. 31 Stylohyoid bone. 32 Ceratohyoid bone. 33 Basihyoid bone. 34 Caudal horn of hyoid (thyrohyoid bone).

Laryngeal and tracheal cartilages

35 Epiglottic cartilage (forming basis of epiglottis). 36 Thyroid cartilage. 37 Cricoid cartilage. 38 Arytenoid cartilage. 39 Tracheal cartilage (one of approximately 35 ring-shaped cartilages, incomplete dorsally).

Vertebral bones

40 Dorsal arch of atlas vertebra (C1). 41 Ventral arch of atlas. 42 Spinous process of axis vertebra (C2). 43 Vertebral body of axis. 44 Odontoid process (dens) of axis (projecting forwards into vertebral canal of atlas). 45 Vertebral (spinal) canal (for housing spinal cord – shown in surface projection in E).

Nasal cavity and paranasal sinuses

46 External nostril (leading into nasal vestibule). 47 Nasal vestibule (surrounded by nasal cartilages). 48 Internasal septum (dividing nasal cavity into left and right nasal fossae). 49 Nasal cavity (shown in surface projection in A & C). 50–53 Air passages (meatuses) through nasal cavity. 50 Dorsal nasal meatus. 51 Middle nasal meatus. 52 Ventral nasal meatus. 53 Nasopharyngeal meatus (airway leading back to internal nostrils). 54 Internal nostrils (leading into nasopharynx). 55 Maxillary recess (lateral diverticulum from nasal cavity shown in surface projection in A, B & C). 56 Nasal opening into maxillary recess. 57 Frontal sinus (shown in surface projection in A, B & C). 58 Lateral part of frontal sinus (large empty space in frontal bone). 59 Medial part of frontal sinus (containing extensions from ethmoidal labyrinth).

Muscles

60 Mylohyoid muscle. 61 Geniohyoid muscle. 62 Sternohyoid muscle. 63 Genioglossal muscle. 64 Intrinsic musculature of tongue. 65 Longus colli muscle. 66 Ventral rectus capitis muscle. 67 Dorsal rectus capitis muscle. 68–69 Semispinal muscle of head. 68 Biventer muscle. 69 Complexus muscle. 70 Dorsal cricoarytenoid muscle (glottic opening muscle).

Oral cavity, pharynx and larynx

71 Oral fissure (mouth opening between lips). 72 Oral cavity (in surface projection in E). 73 Oral vestibule (between lips and cheeks and dental arches and gums). 74 Oral cavity proper. 75–78 Tongue. 75 Apex of tongue. 76 Body of tongue. 77 Root of tongue. 78 Lyssa (rod-like body in free end of tongue). 79 Soft palate (attached to caudal border of hard palate). 80 Palatine veil (soft palate beyond pterygoid hamuli). 81 Position of palatoglossal fold of mucous membrane (between tongue and soft palate and denoting termination of oral cavity caudally). 82 Palatine rugae (transverse ridges of cornified mucosa of hard palate). 83–85 Pharynx (pharyngeal cavity shown in surface projection in E). 83 Nasopharynx (pharynx above soft palate continuing nasopharyngeal meatus of nasal cavity). 84 Oropharynx (pharynx below soft palate continuing oral cavity). 85 Laryngopharynx (common pharynx – confluence of nasopharynx and oropharynx). 86 Palatopharyngeal arch (mucosal fold in lateral pharyngeal wall marking caudal boundary of nasopharynx – arches on either side surround intrapharyngeal opening. 87 Middle ear (tympanic) cavity (contained within tympanic bulla and housing three auditory ossicles shown in surface projection in C). 88 Pharyngotympanic (auditory or Eustachian) tube (linking middle ear cavity with nasopharynx) and pharyngeal opening. 89 Palatine tonsil (in tonsillar fossa). 90 Epiglottis (based on epiglottic cartilage and projecting from pharyngeal floor marking caudal boundary of oropharynx). 91 Aryepiglottic fold (bordering laryngeal vestibule). 92 Laryngeal vestibule. 93 Laryngeal cavity. 94 Trachea (windpipe). 95 Vocal fold (mucous membrane bordering glottis – opening from laryngeal vestibule into larynx proper). 96 Vestibular fold (false vocal fold). 97 Entry into laryngeal ventricle. 98 Oesophagus (gullet).

Central nervous system

99 Cerebral hemisphere of forebrain. 100 Olfactory lobe of forebrain. 101 Corpus callosum (connecting cerebral hemispheres – cut in median plane). 102 Fornix (fibre tract associated with hippocampus). 103 Intermediate mass of thalamus (cut in median plane). 104 Optic chiasma (confluence of right and left optic nerves). 105 Pituitary body (hypophysis). 106 Pineal body. 107 Midbrain. 108 Cerebellum of hindbrain. 109 Pons (cut in median plane). 110 Medulla oblongata. 111 Spinal cord.

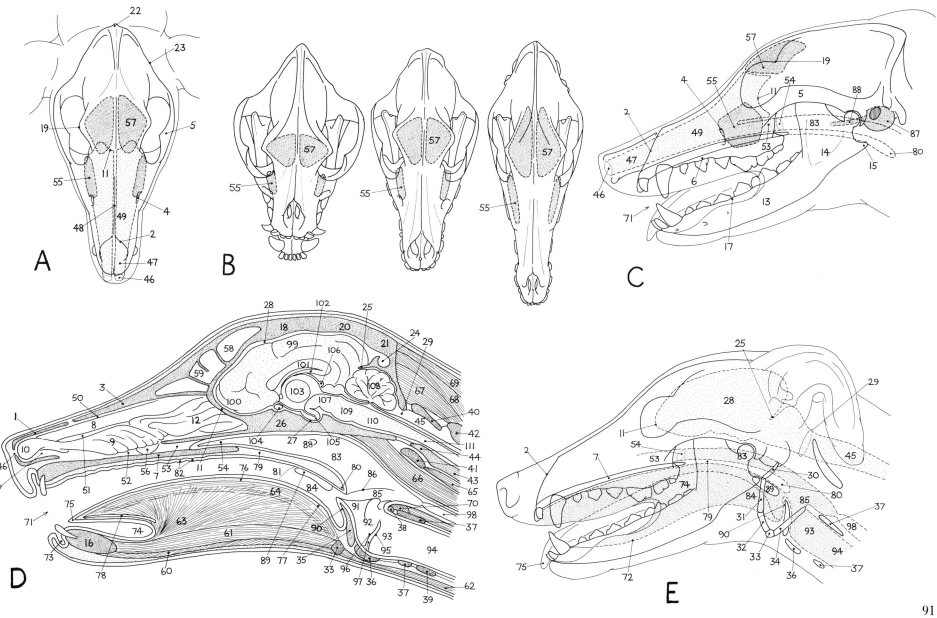

27

THORACIC, ABDOMINAL AND PELVIC VISCERA OF THE BITCH FROM BELOW

These five drawings of the trunk of a bitch lying on her back continue the sequence begun in figs 17 & 18 with progressive removal of various parts of the body to display deeper structures. Again the sequence is preceded by a surface view (**A**) on which the major palpable points and topographical subdivisions are shown.

In drawing **B** the intercostal spaces have been cleared and the abdominal wall has been removed. The lungs are visible *in situ* in the chest but in the abdomen little is visible because the greater omentum covers much of the intestinal mass. The urinary bladder is exposed at the caudal end of the abdomen and a part of the stomach in the xiphoid region at the cranial end. The spleen is exposed in the left hypochondriac region beneath the costal arch.

In drawing **C** most of the ribcage has been removed, the cut stumps of the ribs being left in place flanking the lungs; the greater omentum is removed from its position covering the coils of small intestine; and the diaphragm has been trimmed down to expose the liver divided up by deep fissures into several lobes. The stomach is just visible in this drawing between liver and intestinal mass, with the spleen extending down on its left side towards the ventral midline in the region of the costal arch. The intestinal coils that are now visible, occupying much of the ventral and lateral parts of the belly cavity, are all small intestine (jejunum and ileum).

In drawing **D** the lungs have been removed from the thorax exposing the heart and great vessels within the thoracic cavity and in the base of the neck. In the abdomen the liver has been removed to expose the rear surface of the diaphgram, an exposure which is completed in **E** when the stomach is removed as well. You can now see the peripheral muscular part of the diaphragm inserting into a central tendinous area. This caudal view of the diaphragm also shows the perforations (hiatuses) we mentioned earlier.

In the abdomen the long axis of the **stomach** (indicated by the line of the greater curvature) lies more or less transversely. The *cardia*, through which the oesophagus opens into the stomach, is close to the diaphragm just to the left of the midline: the *pylorus*, leading from the stomach into the duodenum, is over on the right side of the body. An empty stomach is small and contracted towards its cardia a relatively fixed point. It lies cranial to the costal arch wholly within the costal part of the abdomen. This empty stomach is completely hidden from observation and palpation by the liver and diaphragm and its surface and greater curvature are separated from the abdominal floor by the intestines.

As the stomach distends both cardia and pylorus maintain fairly constant positions, the initial enlargement being of the fundus which bulges to the left and caudodorsally coming into contact with the diaphragm and the left lateral abdominal wall and displacing the liver ventrally to some extent.

Further distension enlarges the body of the stomach caudally and ventrally bringing it into contact with the belly wall and left costal arch, pushing the intestines caudally away from any liver contact and pushing the spleen dorsally. A moderately full stomach extends back to a transverse plane through lumbar vertebrae 1 or 2.

A greatly distended stomach extends caudally in the abdominal floor as far as the umbilicus or even in greedy puppies into the caudal abdomen below lumbar vertebrae 3 or 4. The intestines are displaced dorsally and caudally and the liver is pushed onto the right side and the spleen to the left and dorsally.

In drawing **D** most of the small intestine has been removed by cutting through the jejunum just beyond the *duodenojejunal flexure* on the left side, and the ileum just before the *ileocecocolic junction* on the right side. This procedure leaves in place those parts of the gut which we have already specified as keeping a fairly constant position in the abdomen.

Despite its position in the abdominal cavity the **spleen** is not a component of one of the major 'visceral' systems. Its main concern is with the blood system, assisting in the overall formation of blood cells, red and white, and filtering blood and removing and destroying worn out red blood cells. The spleen is attached to the greater curvature of the stomach on the left side of the abdomen. It is an elongated, sickle-shaped organ, anything up to 150 gm in weight and 25 cm in length, lying in the left hypochondriac region following the costal arch ventrally and contacting the flank immediately caudal to the costal arch. The dorsal extremity of the spleen is relatively constant in position below the first lumbar transverse process and the last rib, but its association with the stomach means that the position of its body and ventral extremity will be particularly influenced by the degree of stomach distension. Because of its activity in temporary blood storage spleen size also varies quite considerably which will influence its surface relationships.

When the stomach is empty and the spleen is 'contracted' it extends ventrally in contact with the diaphragm predominantly cranial to the costal arch. As the arch inclines cranially the spleen continues ventrally so that its ventral extremity is exposed on the ventrolateral belly wall caudal to the costal arch.

When the stomach is enlarged and/or the spleen is 'distended' it is pushed caudally beyond the costal arch and its ventral extremity may be as much as twice the width of its dorsal extremity and may extend caudally beyond the unbilicus to a transverse plane through the midlumbar region, and ventrally beyond the midline onto the right side of the belly.

Removal of the remaining parts of the abdominal intestines in **E** exposes structures lying in the abdominal roof. These include **sublumbar muscles** the predominant action of which is flexion of the loins. One of these sublumbar muscles (the *psoas major*) combines with an *iliacus muscle* from the ilium to pass out of the abdomen onto the femur in the thigh. This combined *iliopsoas muscle* promotes hip joint flexion and proves to be one of the major protractors of the hindlimb.

In contact with the sublumbar muscles in the bitch illustrated here are certain components of both the urinary and genital systems. We have already considered at some length the female genital system. The **urinary system** based on paired, bean-shaped kidneys is excretory, removing waste products and excess water from the body as urine. For much of its extent the urinary system is basically the same in dog and bitch. The **kidneys** are the essential excretory organs, waste products to be expelled from the body are removed from blood as it passes through kidney tissue. Blood is conveyed direct to the kidney in a renal branch from the abdominal aorta. For their size relative to other organs the kidneys receive vast quantities of blood. It is estimated that 25% of the output of the heart passes into the kidneys and, moreover, this large blood supply direct from the aorta is under considerable pressure.

Within a kidney an entering renal artery divides into several distributing vessels which in turn branch and re-branch to supply many thousands of minute knots of capillary blood vessels. Because of the high blood pressure that still exists within these capillary complexes, large quantities of fluid containing waste products for excretion are forced out through their walls. However, a complex barrier prevents the passage of components such as blood cells and blood proteins which must remain in the blood. The overall process

is therefore one of filtration under pressure and the kidney filtrate is passed into and along minute, but long and tortuous tubules within the kidney so that water may be reabsorbed if necessary and the urine concentrated. Alternatively water may be added to the filtrate – thus kidneys function in maintaining the water balance of the body.

The kidneys lie in the roof of the abdominal cavity on either side of the aorta and caudal vena cava and beneath the sublumbar muscles. They are approximately equal in size but are asymmetrically placed as is apparent from the drawing (**E**). The right kidney is firmly attached to the abdominal roof beneath the transverse processes of lumbar vertebrae 1−3. Cranially more than half of it lies embedded in the liver, while caudally its ventral surface contacts the descending duodenum. The left kidney lies further back but is more variable in position due to its contact cranially with the stomach. When the stomach is empty the left kidney lies below the transverse processes of lumbar vertebrae 2−4: when the stomach is full it lies further back. Ventrally the left kidney contacts the descending colon (**D**). The lateral surfaces of both kidneys contact the flanks caudal to the costal arch. It is obvious from the drawing that the left kidney will have a greater flank contact than the right and may be located almost midway between last rib and iliac crest. However, the spleen lies against the left flank and may obscure kidney contact.

The tortuous excretory tubules within the substance of a kidney eventually empty urine into a **ureter** which transports it away from the kidney. The

ureter leaves a kidney at its hilus, the indentation in the medial border of its bean-shape, and passes back beneath the sublumbar muscles into the pelvis where it enters the neck of the bladder along with its fellow of the opposite side (the right ureter is shown entering the bladder in **E** in which the bladder has been pulled over to one side). Each ureter is a fibromuscular tube, the muscle moving urine along in a reflex manner. At the bladder a ureter runs for a short distance in the bladder wall before actually opening. As the bladder distends with urine the ureter is collapsed in the wall providing a valve-like action preventing urine from passing back into a ureter when a full bladder contracts. Reflux of urine back up the ureter would exert a pressure on kidney tissue which might prove harmful over a period of time.

The musculomembranous **urinary bladder** (shown in position in **B, C & D**, but moved to the left side in **E**), is a hollow, pear-shaped organ whose size, shape and position vary depending upon the amount of urine inside it. When empty its walls are contracted and therefore thick, and it is a small globular organ lying practically entirely within the pelvis barely protruding into the abdominal cavity. On filling it gradually expands into the abdominal floor pushing the coils of the small intestine forwards, and on extreme distension it rounds off and may extend as far forwards as the navel. The bladder wall contains a large quantity of muscle arranged in a somewhat haphazard manner except at its neck where it forms an encircling sphincter (closure) muscle.

This sphincter, like the internal anal sphincter, relaxes in a reflex manner as the bladder reaches a certain level of distension. However, a dog has control over its bladder because it has a voluntary muscle sphincter surrounding its urethra beyond the neck of the bladder. The dog can relax this voluntary urethral sphincter and urine may flow as and when required.

The neck of the bladder is directly continuous with the urethra, but here dog and bitch differ. A bitch's **urethra** is short and ends within her pelvis opening on a slightly raised urethral tubercle in the floor of the genital tract at the vaginovestibular junction, as we have already noticed and as you can see in fig 29F. This 'external' urethral opening is more or less at the level of the ischiatic arch and from here to the surface of the body at the vulvar cleft a urogenital sinus (formed from the vestibule and vulva) is common to both urinary and genital systems.

In a dog (fig 29C) the urethra is a much longer tube with only that part leading back from the bladder neck into the pelvis being equivalent to the urethra of a bitch. The remaining part of the urethra in the pelvis receives sperm ducts (vasa deferentia) from the testes and continues as a common passageway for both sperm and urine and is therefore comparable to the vestibule of the bitch. The continuation of the urethra into and through the penis to open at its tip at the external urethral orifice has no counterpart in the bitch. The male urethra is therefore a much longer structure and its walls are more heavily muscled. For much of the time its internal cavity is collapsed and the lining of its wall is thrown into numerous longitudinal folds. The lumen is opened up either by urine or the ejaculate being moved along it by muscle activity. In urination movement is predominantly due to bladder contraction, but in ejaculation sperm is moved along by the surrounding urethral muscle, assisted by rhythmic contraction in the bulbospongiosus muscle around the penile bulb.

The large **prostate gland** completely surrounds the pelvic urethra and neck of the bladder at the pelvic inlet, but obviously the degree of bladder distension will effect its position somewhat. A full bladder drags the prostate gland forwards into the abdomen; an empty bladder allows the prostate to lie just inside the pelvic cavity. The sperm ducts pass through the prostate on their way into the pelvic urethra. Because of its position and relationships to urethra, bladder, sperm ducts and rectum (clearly shown in fig 29C), I'm sure you can appreciate how enlargement of the prostate gland in an old dog (not an uncommon happening) may interfere with both urination and particularly defaecation.

Surface features and topographical regions
1 Tracheal region. 2 Presternal region (breast). 3 Sternal region (brisket). 4 Costal (rib) region. 5 Cardiac region. 6 Axillary region (armpit). 7–8 Cranial abdominal (epigastric) region. 7 Right hypochondriac region. 8 Xiphoid region. 9–10 Middle abdominal (hypogastric) region. 9 Right lateral abdominal (iliac) region (flank). 10 Umbilical region. 11 Fold of flank. 12 Umbilicus (navel). 13–14 Caudal abdominal (hypogastric) region. 13 Right inguinal region. 14 Pubic region. 15 Fold of groin. 16 Urogenital (pudendal) region (vulva). 17 Root of tail. 18 Brachial region (arm). 19 Femoral region (thigh). 20 Femoral triangle. 21 Teats of mammary glands.

Bones, joints and ligaments
22 Rib 1. 23 Rib 13 (last or floating rib). 24 Costal arch (fused costal cartilages of ribs 10–12 attached to costal cartilage of rib 9 [last sternal rib]). 25 Sternum (formed from fusion of eight sternebrae – intersternebral cartilages becoming infiltrated with bone). 26 Manubrium of sternum (first sternebra projecting forwards into base of neck). 27 Xiphoid process of sternum (last sternebra extending back into belly wall). 28 Pubic brim (pubic pectens of left and right sides). 29 Ischiatic tuberosity (point of buttock). 30 Ischiatic arch. 31 Pelvic symphysis. 32 Hip joint. 33 Obturator foramen.

Muscles
34 Internal intercostal muscles (beneath parietal pleura). 35–40 Diaphragm (musculotendinous partition between thorax and abdomen). 35 Costal muscle fibres of diaphragm. 36 Crura of diaphragm (arising from lumbar vertebrae). 37 Central tendinous area of diaphragm. 38–40 Diaphragmatic hiatuses (openings for passage of structures between thorax and abdomen). 38 Caval foramen of diaphragm (passage of caudal vena cava). 39 Oesophageal hiatus of diaphragm (passage of oesophagus). 40 Aortic hiatus of diaphragm (passage of aorta). 41 Costal retractor muscle. 42 Quadratus lumborum muscle. 43 Psoas minor muscle. 44–45 Iliopsoas muscle. 44 Psoas major part of iliopsoas muscle. 45 Iliacus part of iliopsoas muscle. 46 Cranial and caudal parts of sartorius muscle. 47 Pectineus muscle. 48 Adductor muscles. 49 Gracilis muscle. 50 Symphyseal tendon (uniting adductor and gracilis muscles of left and right sides and attaching to pelvic symphysis). 51 Semitendinosus muscle of hamstring group. 52 Quadratus femoris muscle. 53 External obturator muscle.

Heart and blood vessels
54–57 Heart. 54 Right ventricle. 55 Left ventricle. 56 Auricular appendage of right atrium. 57 Auricular appendage of left atrium. 58 Aorta (leading from left ventricle). 59 Thoracic aorta. 60 Abdominal aorta. 61 Renal artery to kidney. 62 Left external iliac artery (to hindlimb). 63 Brachiocephalic artery (origin of blood vessels to head, neck and right forelimb). 64 Left common carotid artery (to head). 65 Right subclavian artery. 66 Left subclavian artery. 67 Cranial vena cava (returning blood from front end of body). 68 Right brachiocephalic vein. 69 External jugular vein. 70 Left subclavian vein. 71 Caudal vena cava (returning blood from hind end of body). 72 Renal vein. 73 Pulmonary trunk. 74 Left pulmonary artery.

Internal organs
75 Trachea (windpipe). 76 Tracheal bifurcation (division into left and right principal bronchi). 77 Cranial (apical) lobe of left lung. 78 Middle (cardiac) lobe of left lung. 79 Caudal (diaphragmatic) lobe of left lung. 80 Thymus gland (at its largest in young dog and atrophying with age although never disappearing completely). 81 Oesophagus (gullet). 82–87 Stomach (more or less transversely orientated with bulk on left side). 82 Cardia of stomach (entry from oesophagus). 83 Fundus of stomach. 84 Body of stomach. 85 Pylorus of stomach (exit into duodenum). 86 Greater curvature of stomach (transversely positioned with greater omentum attached to it). 87 Lesser curvature of stomach. 88 Spleen (associated with greater curvature of stomach and costal arch of left side). 89 Greater omentum (enlarged and modified dorsal mesogastrium extensively infiltrated with fat). 90 Liver (moulded on rear surface of diaphragm). 91 Pancreas. 92–97 Small intestine. 92–95 Duodenum (U-shaped loop in abdominal roof). 92 Cranial duodenal flexure. 93 Descending duodenum (against upper part of right flank). 94 Ascending duodenum. 95 Duodenojejunal flexure. 96 Coils of small intestine. 97 Jejunum (cut through just beyond duodenojejunal flexure). 98 Ileum (cut through just before ileocolic junction). 99 Caecum (blind-ending corkscrew-shaped projection from the colon on right side at ileocolic junction). 100–102 Large intestine (colon – a sigmoid shaped structure in 3 parts). 100 Ascending colon (against right flank). 101 Transverse colon. 102 Descending colon (passing back against upper part of left flank). 103 Jejunal lymph nodes. 104 Right kidney. 105 Left kidney (further caudally in abdomen than right). 106 Left ureter (leaving hilus of left kidney). 107 Urinary bladder (receiving ureters from kidneys and temporarily storing urine). 108 Neck of bladder. 109 Urethra (continuation into pelvis of neck of bladder). 110 Right ovary. 111 Right uterine horn. 112 Uterine body (from fusion of uterine horns in midline). 113 Position of uterine cervix. 114 Vagina (continuing uterine cervix into pelvis). 115 Vestibule (urogenital sinus, direct caudal continuation of vagina, entry of urethra marking vaginovestibular junction). 116 Vulva (external genitalia of vulvar cleft surrounded by labia). 117 Left adrenal gland.

Body Cavities
118 Thoracic cavity. 119 Abdominal cavity (containing abdominal part of peritoneal cavity). 120 Pelvic cavity (containing pelvic part of peritoneal cavity opened by removal of midventral sections of pubic and ischiatic bones).

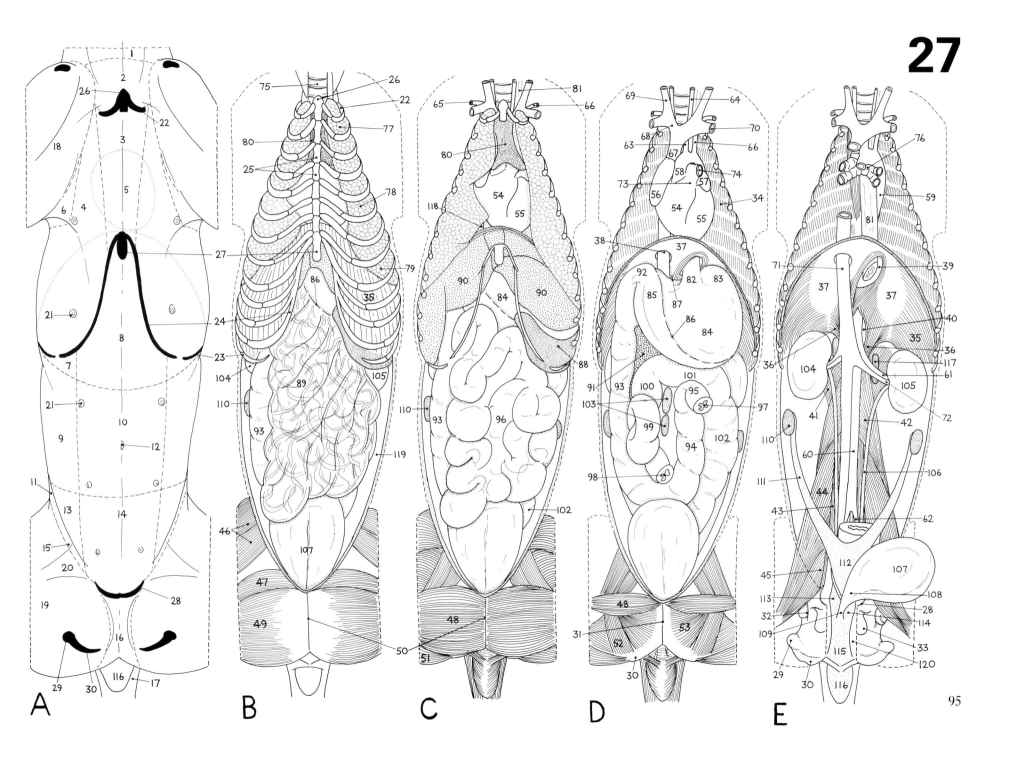

27

95

28

SURFACE PROJECTIONS OF THORACIC, ABDOMINAL AND PELVIC VISCERA OF THE BITCH FROM BELOW

The selection of sketches on this page are outline drawings of the body surface on which have been superimposed the outline of various internal structures to indicate the position they might occupy in relation to the surface. This procedure has already been used from left and right lateral views (figs 23 & 24).

As a starting point **A** shows the body surface on which are indicated the palpable bony points of reference and topographical subdivisions of the trunk. The thorax drawings (**B, C & D**) show initially a projection of the bony limits of the **thoracic cavity** (solid line). However, the more restricted internal longitudinal extent of the cavity is again indicated by a line showing the most cranial extent of the dome of the diaphragm. The outer limits of left and right **pleural cavities** are shown by the broken lines. From **B** it might be assumed that the pleural cavities come close together in the midline of the thorax where they are separated by the narrow connective tissue partition the **mediastinum**. This is in fact the situation in the ventral thorax internal to the sternum where right and left pleural cavities are only separated by a mediastinum composed of a thin sheet of endothoracic fascia supporting the pleura. Because of its thinness the ventral

mediastinum is a potentially fragile structure which might rupture. Further dorsally in the chest, however, the mediastinum is a substantial partition containing the heart, great vessels, trachea and oesophagus and occupying up to one third of the volume inside the thoracic cavity. In **D** the heart is shown in position in the cardiac mediastinum and you can see how the pleural cavities here will be much restricted in size.

Consideration of the thorax is completed in **C** on which approximate positions of the lungs and heart are shown. Surface projections of the lungs and heart again indicate that the lungs are smaller than would be expected. The long axis of the heart is at an angle to the long axis of the body with up to two thirds of the heart being to the left of the median plane.

In **E** the outer boundary of the abdominal and pelvic cavities is shown in the solid line with the outer limit of the peritoneal cavity contained within in a broken line. The retroperitoneal part of the pelvis shown by the heavy stippling is again clearly apparent, the peritoneal cavity not extending very far back through the pelvic inlet so that pelvic viscera are in the main retroperitoneal in position.

In drawings **F** and **G** the approximate positions in surface projection of the principal abdominal and pelvic organs are shown.

Surface features and topographical regions
1 Tracheal region. **2** Lateral neck (jugular) region. **3** Jugular fossa. **4** Presternal (breast) region. **5** Sternal region (brisket). **6** Costal (rib) region. **7** Cardiac region. **8** Axillary region (armpit). **9–10** Cranial abdominal (epigastric)

region. **9** Right hypochondriac region. **10** Xiphoid region. **11–12** Middle abdominal (mesogastric) region. **11** Right lateral abdominal (iliac) region (flank). **12** Umbilical region. **13** Umbilicus (navel). **14** Fold of flank. **15–16** Caudal abdominal (hypogastric) region. **15** Right inguinal region. **16** Pubic region. **17** Fold of groin. **18** Urogenital (pudendal) region. **19** Brachial region (arm). **20** Femoral region (thigh). **21** Femoral triangle.

Bones
22 Rib 1 (at thoracic inlet). **23** Rib 13 (last or floating rib). **24** Costal arch (fused costal cartilages of ribs 10–12 attached to costal cartilage of rib 9 [last sternal rib]). **25** Manubrium of sternum (first sternebra projecting forwards into base of neck). **26** Xiphoid process of sternum (last sternebra extending back into belly wall). **27** Pubic brim (formed from pubic pectens of left and right sides). **28** Ischiatic tuberosity (point of buttock). **29** Ischiatic arch. **30** Greater tubercle of humerus (point of shoulder).

Surface projections of body cavities
31 Surface projection of outer boundary of thoracic cavity – endothoracic fascia lining thoracic cavity (unbroken line including 1st rib [22], costal arch [24] and last rib [23]). **32** Contour of diaphragm at its most cranial extent. **33** Surface projection of outer boundary of abdominal and pelvic cavities (unbroken line including diaphragm [32]). **34** Thoracic inlet (bordered by 1st ribs, thoracic vertebra 1 and sternal manubrium). **35** Pelvic inlet (more or less rounded opening at an oblique angle to long axis of body and bordered above by sacral promontory, below by cranial borders of pubic bones and on either side by arcuate lines on ilia). **36** Pelvic outlet (roughly triangular opening bordered above by caudal vertebra 3, below by ischiatic arch and ischiatic

tuberosities and on either side by sacrotuberous ligaments).

Surface projections of coelomic cavities
37 Surface projection of outer boundaries of left and right pleural cavities – parietal pleura (broken line). **38** Costal pleura. **39** Diaphragmatic pleura. **40** Mediastinal pleura. **41** Costodiaphragmatic line of pleural reflection. **42** Pleural pocket of left pleural cavity. **43** Costomediastinal line of pleural reflection. **44–45** Mediastinum. **44** Ventral mediastinum. **45** Cardiac mediastinum. **46** Surface projection of outer boundary of peritoneal cavity in abdomen and pelvis – parietal peritoneum (broken line). **47** Retroperitoneal component of pelvic cavity. **48–49** Areas available in pleural cavities for lung expansion. **48** Costodiaphragmatic recess. **49** Costomediastinal recess.

Surface projections of thoracic viscera
50 Surface projection of lungs at maximal expiration. **51** Surface projection of heart. **52** Apex of heart.

Surface projections of abdominal and pelvic viscera
53 Fundus of stomach. **54** Greater curvature of stomach. **55** Body of stomach. **56** Pylorus of stomach. **57** Cranial duodenal flexure. **58** Descending duodenum. **59** Duodenal loop. **60** Ascending duodenum. **61** Duodenojejunal flexure. **62** Ileocolic junction. **63** Caecum. **64** Ascending colon. **65** Transverse colon. **66** Descending colon. **67** Rectum. **68** Liver. **69** Spleen. **70** Left kidney. **71** Right kidney. **72** Left ureter. **73** Urinary bladder. **74** Urethra. **75** Left ovary. **76** Right uterine horn. **77** Uterine body. **78** Uterine cervix. **79** Vagina. **80** Vestibule. **81** Vulva.

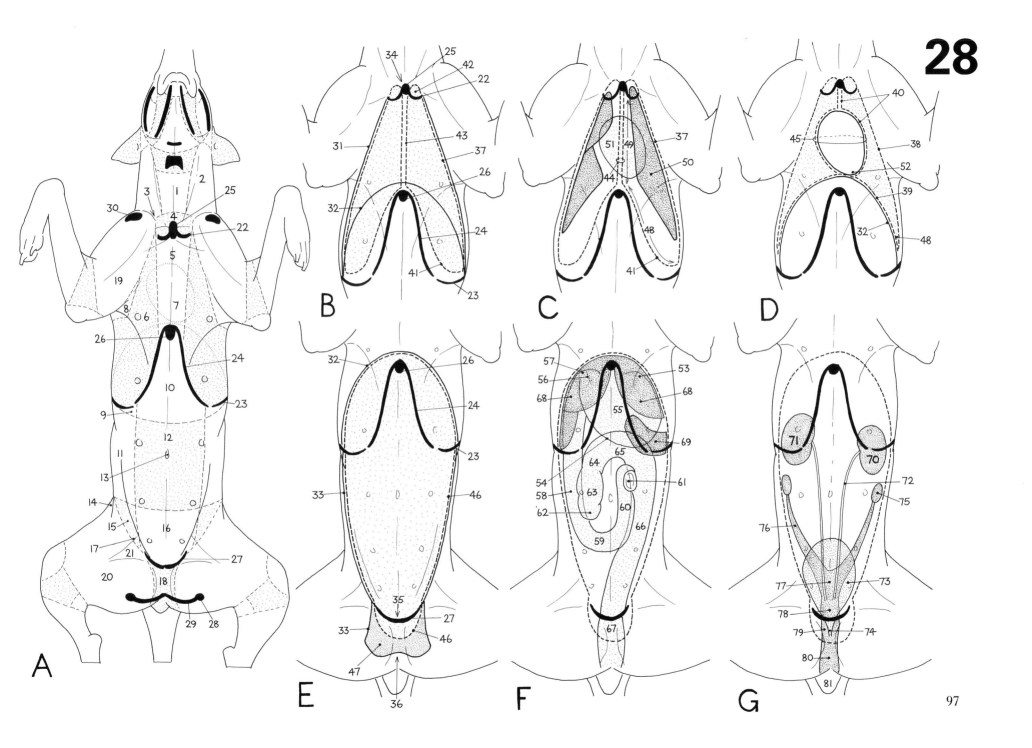

29

SURFACE FEATURES, MUSCLES
AND GENITALIA OF THE DOG
AND BITCH FROM BEHIND AND
IN MIDLINE SECTION

Rear views of the dog and bitch with their tails raised display the perineal region. This can be subdivided by visualizing a horizontal line joining the ischiatic tuberosities, into a dorsal anal region, which borders on the ischiorectal fossae, and a ventral urogenital region which includes the vulva in the bitch and the penile bulb and scrotum in the dog.

The **perineal region** is the representation at the surface of the **perineum**. This is what we have so far talked of as the retroperitoneal part of the pelvis - that part of the body wall that closes (fills) the pelvic outlet. Strictly the perineum will therefore include the muscles and fibrous structures shown in the subcutaneous views, as well as the termination of the pelvic part of the urogenital and digestive tracts. The deep boundaries of the perineum are those of the pelvic outlet: dorsally, caudal vertebra 3; bilaterally the sacrotuberous ligaments and ischiatic tuberosities; ventrally the ischiatic arch.

Drawings **B & E** show the components of the **pelvic diaphragm** which effectively closes off the pelvic cavity about the anus and dorsal perineum. The part of the pelvic cavity lateral to the diaphragm and medial to the sacrotuberous ligament is the **ischiorectal**

fossa. The pelvic diaphragm is composed of symmetrically disposed sheets of striated muscle (*levator ani* and *coccygeus muscles* of left and right sides) running practically parallel to the rectum and urogenital tract, attached laterally to the bony pelvic wall and passing caudomedially and dorsally to close around the anal canal and attach to the underside of the root of the tail.

The pelvic diaphragm is completed caudally by an '**anal diaphragm**' of *external anal sphincter* and *rectococcygeus muscles* around the anal canal in the dorsal perineum, and a '**urogenital diaphragm**' in the ventral perineum. In the dog this latter consists of *bulbospongiosus, ischiocavernosus* and *retractor penis muscles*, principally concerned with erection and ejaculation. In the bitch the bulbospongiosus is represented as *constrictor muscles of the vestibule and vulva* of some importance since they are attached to the ischiatic arch and surround the vestibule to blend with the ventral edge of the external anal sphincter and the pelvic diaphragm. Consequently they help to anchor the reproductive tract against forward drag from a pregnant uterus sinking in the abdomen, and against backward displacement during parturition. They also have some significance in maintaining the 'tie' during copulation.

We have already noticed how the pelvic diaphragm maintains anal position during defaecation through its attachment to the external anal sphincter muscle. Since it is positioned where the pelvic wall is at its weakest, bordering the ischiorectal fossa, it will also assist in retaining viscera within the pelvis whenever intraabdominal and therefore

intrapelvic pressure is raised as in defaecation, micturition and parturition. Essentially it opposes the tendency for pelvic organs to be displaced caudally. Nevertheless displacement might occur in which case a perineal herniation might be palpable as a swelling lateral to the anus.

The diagrams on the right (**C & F**) show the hind part of the abdomen and pelvis approximately in the median plane of the body. However both diagrams also include structures which are positioned away from the midline. In the dog (**C**) the left testis is shown within the scrotum, its duct and blood vessels lateral to the body of the penis within a diagrammatic vaginal process. In the bitch (**F**) the left ovary is shown in the abdomen (it actually lies against the left flank). Leading from it is the left uterine tube continuous with the left uterine horn which passes caudally and medially to meet the right horn in the midline uterine body. Lateral to the left uterine horn the round ligament of the uterus is shown passing lateral to the bladder and entering a subcutaneous vaginal process through an inguinal canal.

The reproductive tracts of dog and bitch have already been described, and you will I'm sure see from these diagrams that there are several comparable features between them. Both have paired gonads in which the sex cells (gametes) form. In a dog testes produce spermatozoa in enormous numbers: in a bitch ovaries produce a much more restricted number of ova. Gonads are also glandular structures which produce sex hormones having considerable effects on the structure of the animal (second-

ary sexual characteristics) and on its behaviour patterns. Both also have paired transportation tubes for conveying the gametes away from the gonads. In functional terms the male duct system remains comparatively simple, transferring sperm from testis to body surface. However, since fertilization takes place internally the duct system of the bitch not only conducts ova away from her ovary but also conveys sperm into her body. The female duct is therefore modified to provide both a site in which internal fertilization may take place (uterine tube), and an area for reception of the penis during copulation (vestibule and vagina), an activity which is a prerequisite for internal fertilization.

Retention of the developing embryo inside the female body has also led to further modifications of the duct system to provide an area for housing, protecting and nourishing the developing foetuses, the uterus. The passage already modified for copulatory purposes (vestibule and vagina) will in turn provide the canal through which foetuses are expelled at birth.

The structure of the **penis** is specifically designed to facilitate internal fertilization. Internally it contains *erectile tissue* composed of numerous blood spaces which can be engorged with blood to expand it. However, expansion is limited by a thick coat of fibrous tissue around the periphery of the penis. An increased blood flow into the organ will enlarge it considerably but, more importantly, will increase its rigidity.

One mass of erectile tissue called the *spongy body* surrounds the whole length of the urethra, with a pair of erectile masses, the *cavernous bodies*, lying above it. Cavernous bodies are considerably more fibrous in nature than the spongy and hence their degree of distension is more restricted. They arise in the penile root from right and left attachments on the ischiatic arch medial to the ischiatic tuberosities. In **C** the left attachment has been severed from the ischiatic arch since half of the pelvic girdle has been removed: in **B** the divergent left and right attachments originating from the ischiatic arch and coming together in the midline are covered by ischiocavernosus muscles. The paired cavernous bodies join and continue along the dorsal surface of the spongy body through penile body and glans to its tip.

Both cavernous and spongy bodies have modifications. The spongy body is expanded proximally as the **penile (urethral) bulb**, in the penile root between the diverging cavernous bodies. More distally the spongy body expands to produce the **erectile body of the glans**, completely surrounding both cavernous and spongy bodies at the free end of the penis. Furthermore it has a distinctive rounded expansion in a dog, the **bulb of the glans**, about midway along the length of the penis.

The paired cavernous bodies join together within the glans penis and become ossified, converted into bone! The glans penis is therefore traversed by a length of bone, the **os penis**, which maintains it in a permanently stiffened state. The significance of such modifications will become apparent when the mechanics of copulation are considered below.

Several muscles are associated with the penis. The **bulbospongiosus muscle**, a continuation of the urethral muscle surrounding the pelvic urethra, covers the urethral bulb in the penile root and the initial part of the urethra in the penile body. Alternative names for this muscle include the accelerator urinae or the ejaculator muscle, both of which give some indications of its actions. A pair of **ischiocavernosus muscles** originate on the ischiatic arch and attach to the fibrous coat of the penile body each one covering one of the diverging crura of the cavernous bodies in the penile root. They assist in anchoring the penis to the pelvis and, during erection, pull the penile root against the ischiatic arch compressing the veins draining blood from the penis. This latter action slows blood flow from the penis helping to maintain the erection. The **penile part of the retractor penis muscle** passes along the underside of the penis to insert onto the glans and into the prepuce. It does not seem to be of much importance in a dog, possibly assisting in rolling back the sheath during the preliminary stage of erection.

Muscle and erectile tissue are also present in the bitch. A **vestibular constrictor muscle** in the vestibular wall is continued as a **constrictor muscle in the vulval lips**. Together these constrictors seem to be equivalent to the urethral and bulbospongiosus muscles of the male. In the lateral walls of the vestibule patches of erectile tissue known as vestibular bulbs are also considered to be equivalent structures to the penile bulb.

Sufficient anatomical information has now been assembled for the overall 'mechanics' of copulation to be outlined. A dog receiving sensory stimulation (predominantly olfactory but also both visual and tactile) from a bitch on heat will become excited and may undergo the first stages of erection. His blood pressure rises and so the flow of blood into the erectile tissue of his penis is increased. At the same time blood flow out of the penis is reduced because the swelling of erectile tissues tends to collapse penile veins through the pressure exerted on their walls. These preliminary stages produce an initial stiffening and elongation of the root and proximal part of the penile body. Remember that the glans penis is *normally* held rigid because of the os penis inside it and so the stiffening and elongation of the root and body of the penis will begin to protrude the glans forwards from the sheath and downwards away from the underside of the belly. As this happens the sheath is gradually rolled back onto the elongating penile body. With continued engorgement of the erectile tissues the root and body of the penis increase in stiffness, the penis protrudes from the sheath even further, and the dog is ready to mount the bitch.

If the bitch is in a receptive mood she will be fairly submissive, presenting her hindquarters to the dog and drawing her tail aside to expose her external genitalia. This pudendal area will be protruded somewhat by contraction in the vulval constrictor muscles which lift the vulval lips whilst at the same time bringing the vestibule into a more horizontal position allowing easier penetration. You will see from **F** how the vestibule normally 'hangs down' over the ischiatic arch. The dog now mounts the bitch, the insertion of his penis into her vestibule and vagina completes the rolling back of the prepuce to expose the entire glans surface.

Once inside the bitch a second stage of erection takes place involving the glans. Up to now this part of the penis has been kept stiff because of the bone inside it. However, a continued increase in blood flow into the penis, coupled with an increased restriction to blood drainage out from the penis, leads to the channelling of blood into the glans and its consequent expansion. A significant flow of venous blood leaving the glans penis actually exits through veins in the prepuce to drain into the external pudendal vein (see fig 37) on the underside of the belly wall. Rolling back of the sheath will effectively block off this channel damming up blood in the glans and promoting its further expansion. The engorgement of erectile tissues is particularly apparent in the bulb of the glans midway along the length of the penis which expands inside the vestibule and vulva locking the dog's penis inside the bitch, 'tieing' the two together. The locking is further enhanced by contraction of the constrictor muscle in the wall of the bitch's vestibule, and by swelling of the erectile tissue in her vestibular walls. Both of these structures clamp around the body of the penis behind the bulbous expansion of the glans and the tie is complete. It is at this stage that the dog may 'dismount' from the bitch turning through 180°. The two remain tied rump to rump and facing in opposite directions.

Ejaculation of sperm bearing semen apparently takes place during the initial phase of copulation, very early in service before the tie fully forms. It would seem that the formation of a tie is not necessarily an essential component of successful copulation. Nevertheless a tie seems to be a normal component of copulation and may last for some considerable time, up to 30 minutes, during which the dog may continue to ejaculate. It has been suggested that the ejaculate which is passed during the tie consists mainly of prostatic gland secretion and this aids in flushing the sperm loaded initial ejaculate further into the bitch.

Surface features and topographical regions
1 Caudal region (tail). **2** Gluteal region (rump). **3** Femoral region (thigh). **4** Ischiatic tuberosity region. **5–8** Perineal region. **5** Anal region. **6** Urogenital region. **7** Pudendal region. **8** Scrotal region (scrotum containing testes). **9** Scrotal raphe (surface representation of internal scrotal division). **10** Ischiorectal fossa (clunial region). **11** Anus. **12** Circumanal skin. **13** Urethral (penile) bulb (expansion of spongy body in penile root between diverging crura). **14** Vulvar cleft. **15** Labia (vulval lips). **16** Inguinal mammary gland.

Bones
17 Lumbar vertebra 3. **18** Wing of sacrum. **19** Lateral sacral crest. **20** Ischiatic tuberosity of hip bone (point of buttock). **21** Ischiatic arch of pelvic girdle (origin for ischiocavernosus and ischiourethral muscles and fibrous cavernous bodies of penis). **22** Pelvic symphysis (cut through in median plane).

Muscles
23 Symphyseal tendon (fibrous plate in median plane attached to pelvic symphysis providing attachment for medial thigh muscles). **24** Superficial gluteal muscle. **25** Gracilis muscle. **26–28** Hamstring muscles.

26 Biceps femoris muscle. **27** Semitendinosus muscle. **28** Semimembranosus muscle. **29** Tail depressors (ventral sacrocaudal muscles). **30** Lateral tail flexors (caudal intertransverse muscles). **31** Tail levators (dorsal sacrocaudal muscles). **32** Internal obturator muscle. **33–34** Pelvic diaphragm. **33** Coccygeus muscle. **34** Levator ani muscle. **35–36** Anal diaphragm. **35** Rectococcygeus muscle. **36** External anal sphincter muscle. **37–41** Urogenital diaphragm. **37** Bulbospongiosus muscle (modified urethral muscle around penile bulb). **38** Constrictor muscle of vestibule (female equivalent of urethral and bulbospongiosus). **39** Constrictor muscle of vulva (labial muscle as for vestibular constrictor). **40** Retractor penis muscle (anal part extends from caudal vertebrae and inserts into rectum and external anal sphincter before passing along underside of penis to preputial fornix). **41** Anal part of retractor penis/clitoridis muscle. **42** Ischiocavernosus muscle (originating from ischiatic arch covering cavernous body in penile root). **43** Ischiourethral muscle. **44** Sacrotuberous ligament (joining ischiatic tuberosity with sacrum and caudal vertebra 1).

Body cavities and peritoneum
45–46 Peritoneum (fluid secreting layer lining abdominal and pelvic cavities). **45** Parietal peritoneum (lining walls of abdominal and pelvic cavities, enclosing peritoneal cavity). **46** Visceral peritoneum (covering abdominal and pelvic viscera). **47** Abdominal part of peritoneal cavity. **48** Pelvic part of peritoneal cavity. **49–52** Pelvic peritoneal excavations. **49** Pubovesical pouch. **50** Vesicogenital pouch. **51** Rectogenital pouch. **52** Pararectal fossa. **53** Retroperitoneal part of pelvic cavity (perineum). **54** Transverse fascia (lining abdominal wall limiting abdominal cavity). **55** Vaginal process (extension of peritoneal cavity

through inguinal canal). **56** Spermatic fascia (enveloping vaginal process – continuation of transverse fascia through inguinal canal). **57–59** Vaginal tunic (extension of parietal peritoneum enclosing cavity of vaginal process). **57** Proper (visceral) vaginal tunic (broken line – covering testis and epididymis). **58** Common (parietal) vaginal tunic (broken line – lining scrotal cavity). **59** Cavity of vaginal process. **60** Vaginal ring (brim of parietal peritoneum surrounding entry into vaginal sac at internal opening of inguinal canal).

Internal organs
61 Rectum (continuation of descending colon through pelvis). **62** Left ureter. **63** Urinary bladder (receiving ureters from kidneys and temporarily storing urine). **64** Urethra (continuation of bladder neck). **65** Pelvic urethra of dog (urogenital sinus – equivalent to vestibule of bitch). **66** Penile urethra of dog (without counterpart in bitch – extends from pelvic outlet to free end of penis). **67** External urethral opening (in floor of vestibule in bitch at vaginovestibular junction and at free end of penis in dog). **68** Testis (spermatozoa produced continuously). **69–71** Epididymis (testicular appendage containing long coiled epididymal duct). **69** Head of epididymis (receiving sperm from testis). **70** Body of epididymis (storing and concentrating sperm). **71** Tail of epididymis (giving origin to vas deferens). **72** Spermatic cord (aggregate of blood vessels and nerves to testis and vas deferens). **73** Spermatic (testicular) artery vein and nerve (passing to/from vessels in abdominal roof). **74** Vas deferens (conveying sperm to pelvic urethra). **75** Prostate gland (accessory sex gland surrounding pelvic urethra at entry of vasa deferentia in pelvic inlet). **76** Cavernous body of penis (pair of fibrous erectile bodies extending through penis from attachments bilaterally on ischiatic arch).

77 Os penis (ossification of both cavernous bodies in glans penis). **78** Spongy body of penis (erectile tissue surrounding penile urethra – continuation of urethral spongy tissue around pelvic urethra). **79** Urethral (penile) bulb (expansion of spongy erectile tissue in penile root between diverging penile crura). **80–81** Glans penis (expansion from spongy erectile body along length of os penis). **80** Bulb of the glans (expansion of spongy erectile body at base of glans). **81** Long part of the glans (elongated expansion of spongy erectile body). **82–83** Prepuce (sheath – enclosing glans). **82** Parietal (middle) layer of sheath. **83** Visceral (inner) layer of sheath (sensory covering of glans penis continuous with urethral epithelium at external urethral opening). **84** Preputial fornix (reflection of middle layer of sheath onto glans surface as inner layer). **85** Preputial cavity (only present in non-erect state of penis, during erection sheath is rolled back along penile body). **86** Preputial opening (leading into preputial cavity). **87** Left ovary (suspended by mesentery from left flank). **88** Ovarian bursa (enclosed by mesentery supporting ovary and uterine tube and into which ova are shed from ovary). **89** Left uterine (Fallopian) tube encircling ovary. **90** Infundibulum of uterine tube (inflated, flared opening leading off from ovarian bursa). **91** Left uterine horn (long and narrow). **92** Uterine body (short from fusion of left and right horns). **93** Uterine cervix (neck, terminating body). **94** Cervical canal (constricted passage through cervix). **95** Round ligament of uterus (assisting in anchorage of uterine horn). **96** Vagina (continuing cervix into pelvis). **97** Vestibule (direct continuation of vagina, urethral entry marks vaginovestibular boundary). **98** Clitoris (female equivalent of male penis lying in vulval floor).

(C & F After Evans & Christensen, 1979).

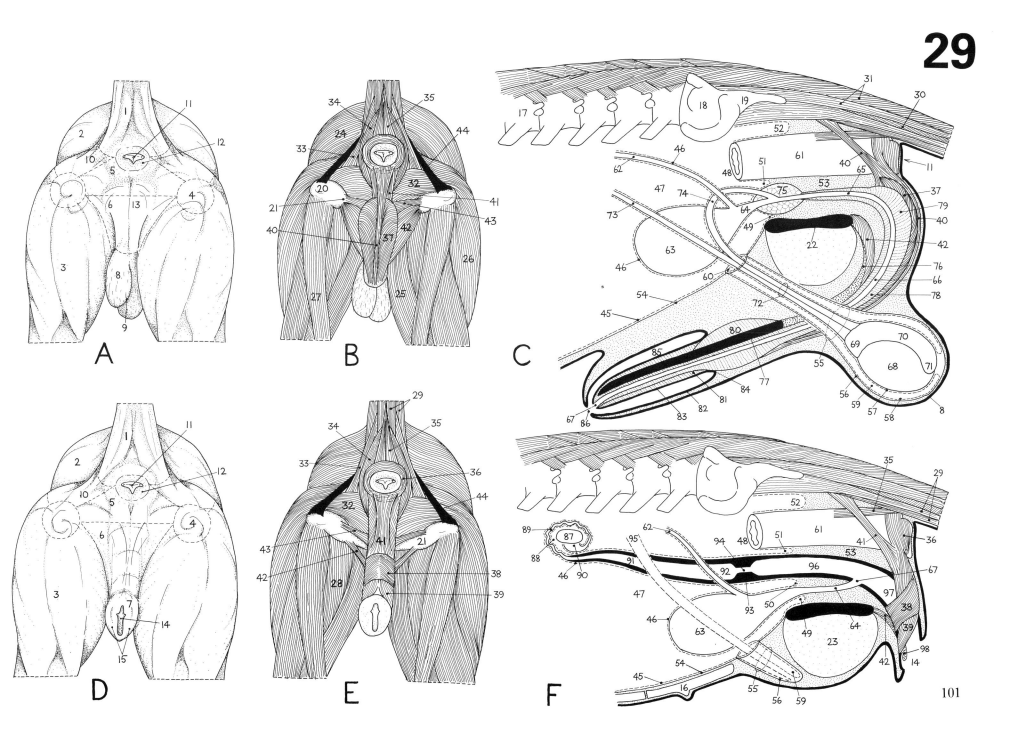

30

BLOOD VESSELS OF THE DOG – ARTERIES

This schema shows the major systemic arteries from the left side. The term **systemic** refers to the blood supply to the body derived from the main aorta. A second circuit, the **pulmonary**, is concerned with circulation through the lungs. By no means all of the systemic arteries are included since this would have made the diagram even more complex than it already is, and would have seriously reduced its usefulness. The diagram therefore leaves out many vessels, for instance the arterial supply to the brain inside the skull deriving from the internal carotid and vertebral arteries, leaving only those vessels to the more superficial head structures deriving from the external carotid artery. The drawing also tries not to duplicate arteries of left and right sides, so for the most part when arteries are symmetrically disposed only arteries of the left side are drawn.

Cell and tissue survival depends upon their proximity to intercellular/tissue fluid. Initially this prevents them from drying out but also provides the medium from which cells obtain nutrients, and into which they pass their waste. These last two factors mean that the fluid must be removed and replenished, it cannot be allowed to stand stagnant. A circulation of fluid thus takes place, and to render it more efficient much of the

system of fluid spaces around and between cells has become enclosed as a complex of transport tubes – the **blood vessels**. Blood plasma is therefore essentially a continuation of tissue fluid even though a cellular boundary, the blood vessel wall, separates the two. Blood corpuscles, red and white, are the cellular component of blood as a tissue, the plasma is the intercellular fluid.

In addition to distributing throughout body tissues the essential components for preserving life, a circulation of body fluids also provides a means of distributing chemical substances (hormones) which help in coordinating and controlling many body activities, as well as being a heat transporter and a defender of the body against disease. Tissue fluid is also siphoned off into a separate system of transport tubes forming a lymphatic system, so that **lymph** is a third type of body fluid together with tissue fluid and blood plasma. More minor but nevertheless important, types of body fluid are also present in specific parts of the body, and several of these we have already alluded to – **synovial fluid** is present in the cavities of the mobile joints: **cerebrospinal fluid** occupies spaces in and around the tissues of the central nervous system (brain and spinal cord): **coelomic fluid** is present as a very thin fluid film in body cavities such as the pleural and peritoneal: **sense organ fluids** such as optic humours in the eyeball and fluid in the inner ear are important in the functioning of the organs of special sensation. All of these fluids are in constant circulation through the intermediaries of blood and lymph systems.

The blood system is a closed network of channels arising from, and eventually returning to, a central pumping station, the **heart**. Vessels leading away from the heart are arteries, those returning blood to the heart are veins. Both types of vessel repeatedly branch and rebranch and are in continuity within the tissues where they are microscopic in size and form **capillaries**. Networks of such capillary vessels permeate throughout all tissues, arterial capillaries leading onto venous capillaries. It is through the extraordinarily thin walls of capillary vessels that exchanges can occur between blood plasma and tissue fluid.

An **artery** is a vessel which conveys blood from the heart to the capillary networks. A continual branching and rebranching of major arteries gives rise to ever smaller arteries and then arteriolar vessels leading to capillaries. Large arteries tend to have very elastic walls, the elasticity being important in two main ways. Firstly, to allow distension of the vessel to accommodate surges of blood when the heart contracts forcing blood through it. Secondly, to recoil elastically and maintain the blood pressure when the heart relaxes. Both of these factors help to smooth out the blood flow entering smaller arteries and thereby reducing pressure fluctuations.

In the walls of smaller arteries muscle increases in amount at the expense of elastic tissue. A more active capacity to adjust their diameter means that such arteries are able to regulate the amount of blood that can flow through them. Blood vessel musculature is of the smooth, involuntary variety so the network has an inbuilt capacity to regulate the flow to large areas of the body in a

reflex manner. This ensures that each tissue receives the amount of blood required to meet its needs. For instance, during exercise the muscle in the walls of small arteries to the limbs will relax to admit an increased blood flow to the limb muscles undergoing activity. Such an increased flow to leg muscles will be at the expense of other areas such as the guts whose arteries contract restricting flow for the time being. It must be borne in mind, nevertheless, that such organs as the brain and heart must have a constant and unfailing supply.

Throughout many parts of the body there are connections (anastomoses) between arterial branches. The possibility is also present for more than one artery being responsible for supplying an area or an organ (**collateral circulation**). In functional terms this will mean that should an artery to a particular area become blocked for any reason, blood will still be able to reach that part through an alternative route. Collateral circuits are particularly noticeable lower down the legs supplementing the supply from a major limb artery. Thus damage or blockage of one vessel need not permanently cut off blood flow to a paw. Within the paws themselves anastomotic connections between vessels gives rise to arterial arches which distribute blood to the digits. A few areas within the body are only supplied by a single artery which lacks communicating links with other arteries or significant collateral channels. Such **end arteries** are common in the heart, brain and kidneys, and if blocked the tissues supplied by them are permanently deprived of blood and will die.

A **vein** is a vessel which conveys

blood back to the heart from capillary networks. A progressive uniting of venules into small veins and then into large veins completes the systemic circuit. Although veins are composed of essentially the same structural components as arteries their walls are thinner and have much less muscle and elastic tissue in them. By the time the blood has passed through a capillary network into venules the pressure producing blood flow, from heart contraction, is much reduced. The structural differences between arteries and veins can be attributed to this difference in pressure — vein walls do not need to be as strong as arterial walls. Low venous pressure does, however, present certain problems for blood return to the heart since it may be counterbalanced or even exceeded by other, external forces such as the force of gravity, particularly noticeable in the limbs. Venous return is also markedly influenced by the contraction of those body muscles bordering veins exerting a squeezing action on them. This external massaging influence is basically non-directional in that it tends to move blood along in either direction in a vein, towards or away from the heart. A one way flow is maintained by valves inside veins. These internal flaps from the vein wall allow flow towards the heart which flattens them against the wall, but they flap open and block the cavity of a vein should any tendency to back flow occur.

Veins also tend to be larger and have a greater capacity than corresponding arteries, while their walls are considerably thinner. Because of their greater capacities veins play a significant role in the storage of blood. Thus at normal blood pressure the volume of blood contained in systemic veins is considerably greater, possibly four or five times as much, as that in systemic arteries. Their thin walls, however, do contain a few muscle fibres which enable them to contract should blood flow decrease and so ensure that a continuous flow is maintained.

The **heart** is the pump within the circulatory system situated in the thorax and so protected to a considerable extent by the surrounding ribcage. In order to perform its rhythmic, contractile functions effectively it must not be embedded in solid tissue which would restrict its movements. We have already seen such a necessity in connection with the guts and the lungs with the presence of pleural and peritoneal cavities and their lining fluid secreting membranes. A similar anatomical arrangement exists in relation to the heart — it is housed in a **pericardial cavity** with a layer of serous membrane covering its surface, and a layer lining the inside of the wall of the cavity. A thin film of pericardial fluid allows friction free unimpeded movement between the two layers.

In order to act as a mechanical pump the heart has walls formed predominantly of muscle tissue continuous with the muscle of those arteries leaving and those veins entering it. Blood enters the two upper cavities of the heart — right and left **atria**, and leaves from the two lower cavities — right and left **ventricles**. The muscle of the atria is separated from that of the ventricles by a complete partition of fibrous connective tissue onto which the heart muscle fibres are attached. This fibrous **cardiac skeleton** is important in two main ways. Firstly, it completely separates atrial from ventricular musculature, the only connection between the two is by a special strand of nervous conducting tissue which extends to the apex of the heart. What this means is that excitation of the heart muscle originating in a 'pacemaker' in the atrial muscle is prevented from stimulating the bases of the ventricles before their apices. Secondly, the skeleton provides a firm framework for support of the fibrous cusps of the valves inside the heart.

Blood enters the **right atrium** in the venae cavae (cranial and caudal) during atrial relaxation. On atrial contraction blood is passed through the **right atrioventricular opening** into the **right ventricle**. The opening contains a valve which closes at the end of atrial contraction preventing regurgitation of blood into the atrium when the ventricle contracts. Right ventricular contraction forces blood into the **pulmonary trunk** through the **pulmonary opening**. From this trunk the pulmonary arteries lead to the lungs. At the end of ventricular contraction the **pulmonary valve** in the pulmonary opening closes preventing regurgitation when the ventricle relaxes.

After permeating through the lungs blood is collected up in the pulmonary veins which empty into the **left atrium** during atrial relaxation. On contraction the oxygenated blood is passed on through the **left atrioventricular opening**, guarded by a valve as on the right side, into the **left ventricle**. From the left ventricle blood is forced by ventricular contraction into the **aorta** through an **aortic opening** guarded by a valve much like the pulmonary valve, and so into the systemic arterial circuit to the body generally.

The activities of left and right sides are synchronized so that atria contract together and ventricles together, while atrioventricular valves open and close together as do pulmonary and aortic valves. Left and right sides differ, however, in the pressures generated on contraction. Since the resistance to blood flow is considerably greater in the systemic circuit than in the pulmonary, the left side must be much stronger, this strength being demonstrated by a greater amount of muscle in the left ventricle.

As you may see from the schema the **aorta** is the main systemic distributing artery. Arising from the left ventricle it extends through thoracic and abdominal roofs to terminate as the **median sacral artery** in the pelvic roof.

The head and neck receive most of their blood in the **common carotid arteries** (left and right) arising in common with the right subclavian artery from the **brachiocephalic trunk**, the first main element arising from the initial part of the aorta, the **aortic arch**. A common carotid divides into an **external carotid artery**, the main continuation of the common supplying extracranial structures of the head, and an **internal carotid artery** entering the cranial cavity as the major supply to the brain. Additional blood flow to the neck and also into the head is in the **vertebral arteries** (right and left) arising from the subclavian arteries, supplying the spinal cord and ultimately entering the cranium to supply much of the hind part of the brain.

The forelimbs receive blood from the **subclavian arteries** (right from the brachiocephalic trunk: left as the second branch from the aortic arch), each continuing across the axilla as the **axillary artery** and into the arm as the **brachial artery**. In the forearm it is continued as the **median artery** and into the palmar surface of the paw where it is distributed as the **palmar common digital arteries** and finally **palmar proper digital arteries**. Forelimb distribution is in a number of branches arising from this main trunk *en route* through the limb.

The hindlimbs receive blood from the **external iliac arteries** (right and left) from the abdominal aorta. An external iliac leaves the abdominal cavity to continue in the thigh as the **femoral artery**. Behind the stifle joint it is continued as the **popliteal artery**, in the shank as the **cranial tibial artery** and into the dorsum of the paw as the **dorsal pedal artery**. It penetrates through to the plantar surface of the paw as the **perforating metatarsal artery** and is distributed to the digits as the **plantar metatarsal arteries** and finally as the **plantar proper digital arteries**. Hindlimb distribution is in a number of branches from the main trunk *en route* through the limb.

The thorax receives blood to its walls primarily through **intercostal arteries** (right and left): dorsal intercostals from the thoracic aorta and ventral intercostals from the subclavian via the **internal thoracic arteries** (right and left). The dorsal intercostals also supply the spinal cord. Blood to thoracic viscera is in the **bronchoesophageal arteries** from the thoracic aorta to the lungs and bronchi, and in the **coronary arteries** (right and left) to the heart from the aorta close to its origin from the left ventricle.

The abdomen receives blood to its walls dorsally through **phrenicoabdominal arteries** (right and left), a series of **lumbar arteries** (right and left) and **deep circumflex iliac arteries** (right and left) all arising from the abdominal aorta: ventrally from **epigastric arteries** (right and left) arising cranially from internal thoracic arteries and caudally from external iliac arteries via **deep femoral arteries** (right and left). The lumbar arteries also supply the spinal cord. Blood to abdominal viscera is from three unpaired branches from the abdominal aorta – **coeliac artery, cranial mesenteric artery** and **caudal mesenteric artery**, and from paired aortic branches – **renal arteries** (right and left) and **gonadial arteries** (right and left).

The pelvis receives blood from the **internal iliac arteries** (right and left) arising from the abdominal aorta caudal to the external iliacs. An internal iliac gives rise to a **caudal gluteal artery** supplying the pelvic wall and rump and terminating in the tail and thigh, and an **internal pudendal artery** supplying the pelvic viscera and external genitalia.

Arterial supply to head, neck, forelimb and thorax
1 Aortic arch. **2** Brachiocephalic trunk. **3** Left common carotid artery. **4** Cranial thyroid and cranial laryngeal arteries. **5** External carotid artery. **6** Occipital artery. **7** Lingual artery. **8** Facial artery with sublingual and dorsal (superior) and ventral (inferior) labial branches. **9** Caudal auricular artery. **10** Superficial temporal artery. **11** Maxillary artery. **12** Mandibular alveolar artery (supplying lower dental arch and with terminal mental branches). **13** External ophthalmic artery. **14** Rostral and caudal deep temporal arteries. **15** Infraorbital artery (supplying upper dental arch and with terminal dorsal and lateral nasal arteries). **16** Internal carotid artery. **17** Carotid sinus. **18** Left subclavian artery. **19** Vertebral artery. **20** Spinal arteries. **21** Anastomotic branch between vertebral and occipital arteries. **22** Costocervical artery. **23** Dorsal scapular artery. **24** Deep cervical artery. **25** Thoracic vertebral artery. **26** Dorsal intercostal arteries. **27** Dorsal branches to epaxial muscles. **28** Internal thoracic (mammary) artery. **29** Pericardiacophrenic artery. **30** Ventral intercostal arteries. **31** Musculophrenic artery. **32** Cranial epigastric artery with superficial and deep branches. **33** Superficial cervical artery with deltoid and ascending branches. **34** Axillary artery (direct continuation of subclavian artery). **35** Subscapular artery (cut through after scapular removal consequently circumflex scapular and thoracodorsal branches to chest wall removed). **36** Caudal circumflex humeral artery. **37** Cranial circumflex humeral artery. **38** Brachial artery (direct continuation of axillary artery). **39** Deep brachial artery. **40** Collateral ulnar artery. **41** Superficial brachial artery. **42** Cranial superficial antebrachial artery (medial and lateral branches). **43** Common interosseous artery with ulnar and recurrent ulnar branches into forearm. **44** Caudal interosseous artery (main continuation of common interosseous). **45** Caudal interosseous artery, dorsal branch. **46** Median artery (direct continuation of brachial artery). **47** Radial artery with dorsal and palmar carpal branches. **48** Dorsal common digital arteries. **49** Dorsal metacarpal arteries. **50** Dorsal proper digital arteries. **51** Superficial palmar arterial arch. **52** Palmar common digital arteries. **53** Palmar metacarpal arteries. **54** Palmar proper digital arteries.

Arterial supply to abdomen, pelvis and hindlimbs
55 Thoracic aorta. **56** Dorsal costoabdominal artery. **57** Abdominal aorta. **58** Phrenicoabdominal artery. **59** Lumbar arteries. **60** Deep circumflex iliac artery. **61** Median sacral artery. **62** Median caudal artery. **63** Caudal arteries. **64** Coeliac artery with gastric, hepatic, pancreaticoduodenal and splenic branches. **65** Cranial mesenteric artery with caudal pancreaticoduodenal, ileocolic and numerous intestinal branches. **66** Caudal mesenteric artery with left colic artery and cranial rectal branches. **67** Left renal artery. **68** Left gonadial (testicular/ovarian) artery. **69** Left internal iliac artery. **70** Caudal gluteal artery. **71** Iliolumbar artery. **72** Cranial gluteal artery. **73** Lateral caudal artery. **74** Internal pudendal artery with urogenital and penile branches. **75** Left external iliac artery. **76** Deep femoral artery. **77** Pudendoepigastric artery. **78** Caudal epigastric artery with deep and superficial branches. **79** External pudendal artery with medial circumflex femoral and obturator branches. **80** Femoral artery (direct continuation of external iliac artery). **81** Lateral circumflex femoral artery. **82** Proximal, middle and distal caudal femoral arteries. **83** Descending genicular artery. **84** Saphenous artery. **85** Cranial branch of saphenous artery. **86** Caudal branch of saphenous artery. **87** Popliteal artery. **88** Cranial tibial artery. **89** Superficial branch of cranial tibial artery. **90** Dorsal pedal artery. **91** Caudal tibial artery. **92** Perforating metatarsal artery. **93** Dorsal common digital arteries. **94** Dorsal metatarsal arteries. **95** Lateral plantar artery. **96** Medial plantar artery. **97** Deep plantar arterial arch. **98** Plantar metatarsal arteries. **99** Plantar proper digital arteries. **100** Pulmonary trunk. **101** Left pulmonary artery. **102** Left ventricle of heart. **103** Left atrium of heart. **104** Pulmonary veins.

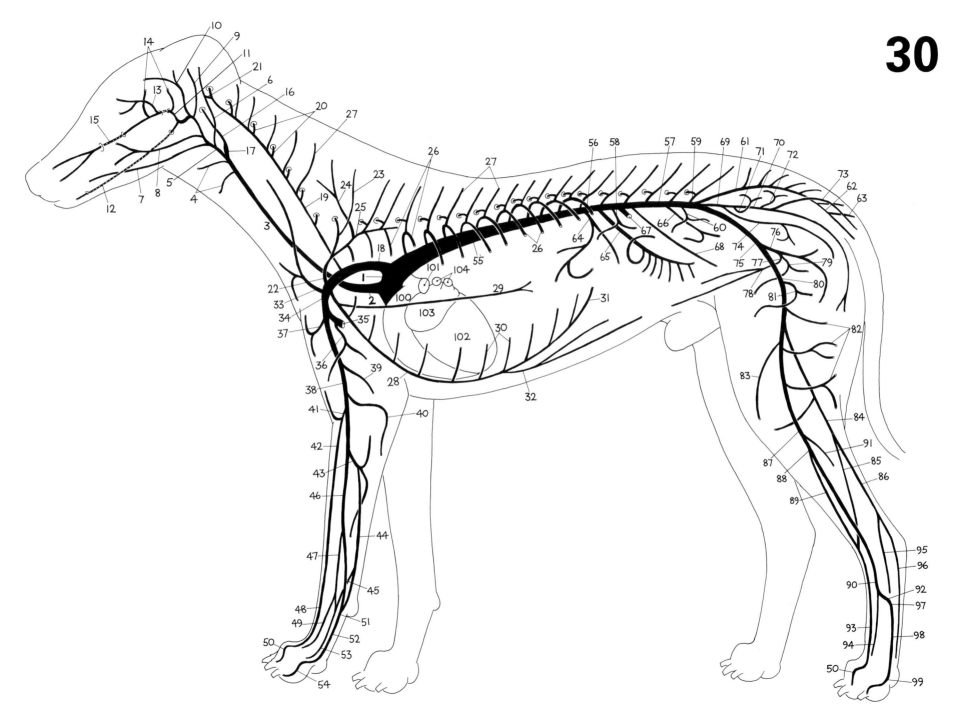

30

31

BLOOD VESSELS OF THE DOG – VEINS

This second blood vessel schema shows the major systemic veins from the right side. As with the arterial diagram by no means all of the veins have been included since the drawing would have become too complex, and when vessels are symmetrically disposed right and left sides, only the right side has been included.

The arrangement and distribution of veins is more variable than that of arteries. This is presumably explained because mechanical as well as inherited genetic factors play a role in the formation of blood pathways in the developing animal. Preferred channels are 'selected' by these factors out of a diffuse vascular network in the early embryo. Since the speed of flow in veins is considerably less than in arteries, mechanical influences will be lessened and the possibility that different pathways will be 'chosen' is thereby increased. The end result is that veins are less regular in arrangement than arteries and they also tend to be more numerous. They often communicate freely with each other and, although generally running with arteries, an accompanying vein may be paired, one on either side of the artery.

Despite some notable exceptions the overall distribution of venous channels parallels arteries to a considerable extent. Thus many arteries and veins run together serving comparable parts of the body and consequently having the same names. However there are a number of exceptions such as the **hepatic portal vein** draining the guts which has no arterial counterpart running with it. It has been drawn in the schema in isolation in the abdomen since it terminates in the sinusoidal networks of the liver. Subsequent drainage from the liver passes into the caudal vena cava through the hepatic veins which are shown as cut stumps in the drawing. In the limbs also we have already seen that the system of deep veins accompanying arteries is supplemented by a superficial, subcutaneous venous circuit. The **cephalic** venous network in the forelimb drains into the axillary vein and external jugular vein; the **saphenous** venous network in the hindlimb drains into the femoral vein.

In the head we see considerable differences between arterial and venous distribution. Drainage from the face and musculature outside the cranium is into tributaries eventually draining into the external jugular vein running down the neck superficially. Drainage from the brain inside the cranium is into a separate system of **venous sinuses** located in the dural layer of the meninges. These pursue a different course to the main arterial supply to the brain in the internal carotid artery. Some of this head drainage is channelled back into the vertebral canal as large **internal vertebral venous plexuses** accompanying the spinal cord. These venous plexuses also receive blood from the epaxial musculature and drain into vertebral, intercostal and lumbar vessels. As we shall see later these plexuses play a significant role in 'cushioning' of the spinal cord.

Venous return from head, neck, forelimbs and thorax

1 Cranial vena cava. 2 Azygos vein. 3 Dorsal intercostal veins. 4 Intervertebral veins. 5 Costoabdominal vein. 6 Lumbar veins. 7 Costocervical vein. 8 Vertebral vein. 9 Dorsal scapular vein. 10 Deep cervical vein. 11 Thoracic vertebral vein. 12 Venous branches from epaxial muscles. 13 Internal thoracic (mammary) vein. 14 Ventral intercostal veins. 15 Musculophrenic vein. 16 Cranial epigastric vein with superficial and deep branches. 17 Brachiocephalic vein. 18 Internal jugular vein. 19 Thyroid vein. 20 External jugular vein. 21 Cephalic vein. 22 Accessory cephalic vein. 23 Median cubital vein. 24 Axillobrachial vein (receiving a lateral thoracic vein from chest wall). 25 Omobrachial vein. 26 Linguofacial vein. 27 Lingual vein with sublingual branch. 28 Facial vein. 29 Deep facial vein. 30 Connection between deep facial vein and external ophthalmic vein. 31 Angular vein of eye (connection between facial vein and external ophthalmic vein). 32 External ophthalmic vein. 33 Maxillary (superior) labial vein. 34 Dorsal and lateral nasal veins. 35 Maxillary vein. 36 Caudal auricular vein. 37 Superficial temporal vein. 38 Mandibular alveolar vein (draining lower dental arch). 39 Dorsal sagittal sinus. 40 Sigmoid sinus. 41 Temporal sinus. 42 Cavernous sinus continued caudally as ventral petrosal sinus. 43 Ventral internal vertebral venous plexus. 44 Subclavian vein. 45 Axillary vein. 46 Subscapular vein (cut through after scapular removal receiving thoracodorsal vein from chest wall and caudal circumflex humeral vein). 47 Brachial vein. 48 Deep brachial vein. 49 Collateral ulnar vein. 50 Common interosseous vein with cranial and caudal interosseous tributaries. 51 Median vein. 52 Deep (palmar) antebrachial vein. 53 Radial vein. 54 Dorsal common digital veins. 55 Dorsal proper digital veins. 56 Superficial palmar venous arch. 57 Palmar common digital veins. 58 Palmar proper digital vein.

Venous return from abdomen, pelvis and hindlimbs

59 Caudal vena cava. 60 Phrenicoabdominal vein. 61 Deep circumflex iliac vein. 62 Hepatic veins. 63 Gonadial (testicular/ovarian) vein. 64 Renal vein. 65 Hepatic portal vein with gastroduodenal, pancreaticoduodenal, splenic, cranial mesenteric, intestinal, and caudal mesenteric tributaries. 66 Right common iliac vein. 67 Median sacral vein. 68 Sacral veins. 69 Median caudal vein. 70 Internal iliac vein. 71 Urogenital vein. 72 Cranial gluteal vein. 73 Lateral caudal vein. 74 Caudal gluteal vein. 75 Internal pudendal vein. 76 Right external iliac vein. 77 Deep femoral vein. 78 Medial circumflex femoral vein. 79 Pudendoepigastric vein. 80 Caudal epigastric vein with superficial and deep branches. 81 External pudendal vein. 82 Femoral vein. 83 Lateral circumflex femoral vein. 84 Proximal, middle and caudal femoral veins. 85 Medial saphenous vein. 86 Cranial branch of medial saphenous vein. 87 Caudal branch of medial saphenous vein. 88 Descending genicular vein. 89 Lateral saphenous vein. 90 Cranial branch of lateral saphenous vein. 91 Caudal branches of lateral saphenous vein. 92 Popliteal vein. 93 Cranial tibial vein. 94 Dorsal pedal vein. 95 Caudal tibial vein. 96 Perforating metatarsal vein. 97 Dorsal common digital veins. 98 Dorsal proper digital veins. 99 Plantar common digital veins. 100 Plantar proper digital veins. 101 Right atrium of heart. 102 Right ventricle of heart.

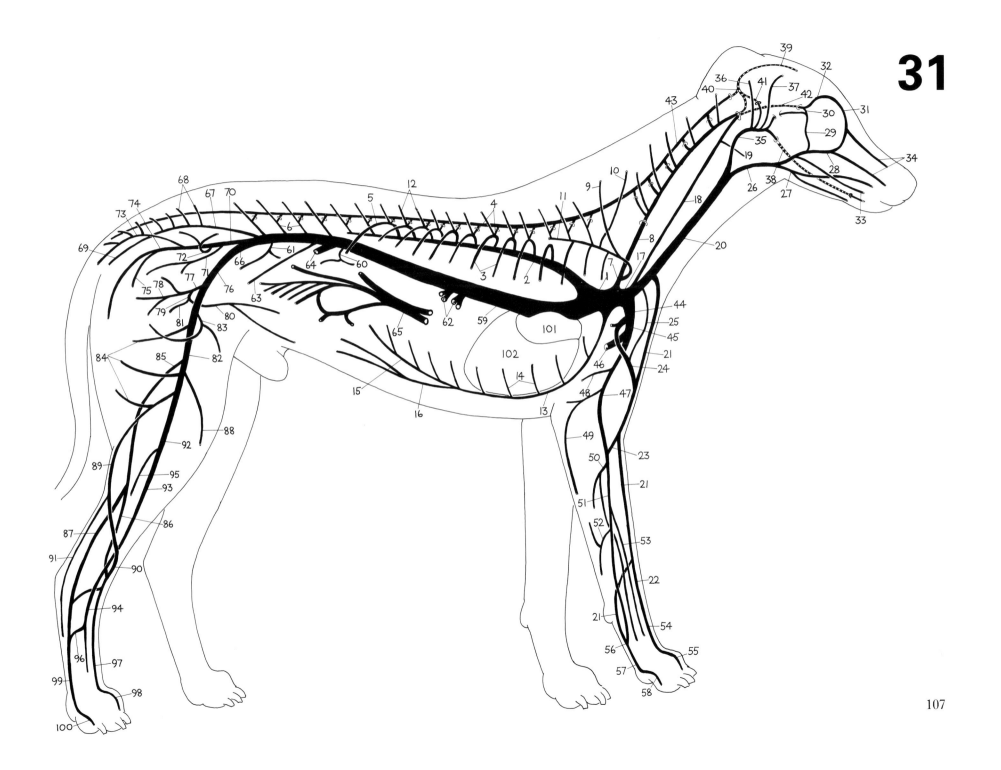

31

107

32

LYMPH DRAINAGE OF THE DOG AND THE POSITION OF THE MAIN LYMPH NODES

The lymphatic system basically represents an additional mechanism for returning fluids from tissue spaces to the blood system; ie. in addition to the venous system. In its passage through blood capillary networks a considerable amount of the fluid component of blood diffuses out through the walls of arterial capillaries into the surrounding tissue spaces. This outward movement of fluid, due in part to the blood pressure set up by pumping of the heart, conveys oxygen and food materials into tissue spaces from where they can pass into cells. At the same time waste products will be passed out of cells into tissue fluid and will subsequently be reabsorbed from the tissue spaces into the venous side of the capillary network. This fluid reabsorption is assisted by the presence of complex blood plasma proteins which do not easily escape through arterial capillary walls. Tissue fluid is therefore a very weak solution of protein. In its passage through a capillary network, as fluid is lost, blood proteins become more concentrated and exert an attractive influence (osmotic pressure) in the venous capillaries drawing fluid back in from the tissue spaces. Fluid reabsorption is also assisted by the significant drop in blood pressure that occurs through the capillary network. Nevertheless *slightly more*

fluid passes out of arterial capillaries than reenters venous capillaries, so that tissue fluid is being continually added to, potentially increasing its overall volume. In addition, some of the complex blood proteins that do by chance pass out into tissue spaces cannot be reabsorbed back into venous capillaries and so remain in tissue fluid. This continual accumulation of excess tissue fluid and large protein molecules must be siphoned off by some method other than in venous capillaries or tissues will become 'waterlogged'. Hence the necessity for a lymphatic system is explained.

'Waterlogging' (**oedema**) of tissues might result from the blood pressure in arterial capillaries being too high — pushing out too much fluid; or the osmotic pressure in venous capillaries being too low — reabsorbing too little fluid. In either situation the balance between the two processes is disrupted and fluid passes out of the bloodstream in considerably larger quantities than it returns. The relative roles of the blood and lympathic systems in fluid reabsorption seems to be that — blood capillaries reabsorb tissue fluid and those soluble crystalloid materials dissolved in it, while lymphatic capillaries reabsorb tissue fluid and larger particulate matter dispersed through it. As the size of molecules in tissue fluid increases there is a change over from uptake into blood capillaries to uptake into lymphatic capillaries.

The lymphatic system is based on thin walled lymphatic vessels found throughout the body except in the central nervous system, the splenic pulp and bone marrow, and also in articular cartilage. It begins as a network of **lymphatic capillaries** in association with the arteriovenous capillary networks which permeate throughout all body tissues. Lymphatic capillaries have no continuity with blood capillaries they simply end blindly within tissue spaces bathed in tissue fluid. Their walls appear to be capable of quite extensive changes explaining how lymphatic capillaries take up large proteins and also particulate matter together with fluid.

Extensive recombination of lymphatic capillaries produces larger lymphatic channels which show many similarities to veins but have even thinner walls. The thinness of lymphatic vessel walls, together with the practically colourless nature of the contained fluid (now called lymph), means that lymphatic vessels are extraordinarily difficult to identify. It is only the very largest of lymph vessels that can normally be grossly recognized without special treatment of tissues. However, the lymph vessels draining the small intestine are responsible for conveying fat absorbed from the intestine. These channels are called **lacteals** and the fat inside them is visible through their thin walls. They appear as yellowy-white streaks running through the supporting mesentery of the small intestine. Ultimately the largest lymphatic vessels coalesce and channel lymph back into the blood stream close to the heart.

Circulation of lymph through the system is entirely dependent upon sources outside it since there is no blood pressure behind lymph flow. Movement is therefore normally slow and produced by the squeezing or massaging of lymph vessels resulting from such activities as the contraction of adjacent body muscles or the pulsations from neighbouring blood vessels, but may simply be due to gravitational forces. Lymph flow is consequently speeded up during general body activity and exercise whilst prolonged immobility of any particular part of the body slows down both venous and lymphatic drainage from that part with a consequent tendency for tissue fluids to accumulate. In order to ensure a one way lymph flow the lymphatic vessels are provided with valves at frequent intervals similar to those in veins. The lack of pressure behind lymph flow also explains why lymph reentry into the blood stream takes place so close to the heart since it is here that the blood pressure is at its lowest point and unimpeded entry of the low pressure lymph stream can occur.

The capacity of lymphatic vessels to take up particulate matter from tissue spaces can itself present problems to a dog. Should any potentially harmful material be present in tissue fluid then the lymphatic system could provide a ready channel for its spread to other parts of the body. Thus all lymph prior to its reentry into the blood stream passes through at least one **lymph node** in which it can be filtered. Filtration of lymph by a node is to a certain extent a purely mechanical process, lymph percolating through a close-meshed sieve of fibrous tissue within the node which will tend to trap such gross matter as dust or carbon particles. More importantly, however, lymph nodes also have the capacity to filter lymph 'biologically' since special cells located in them have the ability to engulf and destroy harmful material passing through. These cells (macrophages) can take up bacteria, virus particles and other invading material, but can also scavenge such things as bits of broken cells and worn out red blood corpuscles which might be present in lymph.

In addition to being both mechanical and biological filters, lymph nodes also represent one of the components of a widespread defence (immune) system — the **lymphoid system**. This system also includes the *red bone marrow* and such organs as the *thymus gland* in the mediastinum cranial to the heart, and the *spleen* in the abdomen. It also includes more diffuse masses of lymphoid tissue associated with the mucosal lining layers in the body, the *tonsils* for instance in the throat, and nodular masses in the intestinal walls. Within all of these anatomical components of the lymphoid system the dominant cells are a specific type of white blood cell, the **lymphocyte**. The immune system seems to be able to react to 'foreign' protein; ie. protein from outside the body. In a lymph node the unfamiliar protein (*antigen*) is trapped and concentrated by macrophages and is subsequently available for the masses of lymphocytes in the node to act upon. Lymphocytes stimulated by this information may make an immediate response within the node: numerous lymphocytes can, by surrounding the invading organism (be it bacterial or viral), isolate and destroy it through the production of poisonous substances which act directly on the antigen. Lymphocytes may also be involved in a

more widespread reaction to invading antigen. In this process they become transformed into special cells able to produce *antibody* which combines with antigen and renders it harmless. Released into lymph such lymphocytes can circulate throughout the body and produce antibodies which are able to destroy the invading organisms or the poisons they produce. By such methods as these a dog can fight against disease and in fact may often succeed in preventing the spread of an infection.

As you can no doubt see from the drawings lymph drainage from a particular area of the body is filtered through a node or collection of nodes specific to that area. Thus infection arising in that part of the body may be confined to it by the efficient action of these nodes. However, confining the infection may lead to the tissue of the node itself becoming infected; it may swell and become very painful. An explanation as to why lymphoid tissue is scattered throughout the body in relation to the mucosal membranes is that these are areas where infection is most likely to occur.

Lymph nodes tend to be oval or bean-shaped with an indented border. Lymph vessels passing to a node enter through its convex surface, lymph vessels leaving a node exit at its indented concave border. Nodes are normally found embedded in fat on the surface of organs or structures, rather than being contained within them. Their size varies (from minute and barely discernible to diameters of several cms) as often does the number at a particular site (several smaller nodes

taking the place of a single larger one).

Most lymph nodes are deeply embedded inside the body some way from its surface, a feature which the drawings do not adequately convey. A few nodes are more superficially placed, and one or two may even be felt through the skin at the surface. Notable in this last respect are the **mandibular nodes** on each side of the throat behind the angles of the lower jaw, and the **popliteal node** of the hindleg in a deep indentation behind each stifle joint. Other nodes, such as the **superficial cervical** on the side of the neck just in front of the shoulder blade, may be palpable if they are swollen for any reason.

The five drawings shown here present in a somewhat diagrammatic fashion some essential aspects of lymphatic drainage.

Drawing **A** shows those lymph nodes receiving drainage of lymph from the skin and subcutaneous structures — the cutaneous drainage.

Drawings **B, C & E** show lymph nodes receiving drainage from the muscles and bones of the body wall and limbs, the bulk of the body. This 'parietal' drainage also includes lymph from the lining layers of the body cavities, that is from the parietal peritoneum of the abdomen and pelvis, and from the parietal pleurae in the thorax.

Drawing **D** depicts those lymph nodes responsible for the drainage of lymph from viscera within the thorax, abdomen and pelvis, but also draining glandular structures in the head and neck, the tongue, pharynx, larynx, trachea and oesophagus.

Drawing **F** at the right of the page shows you that many of the lymph nodes are symmetrically disposed right and left of the body. It also shows that lymphatic drainage channels display considerable symmetry especially the larger channels such as the **lumbar lymph trunks** — paired trunks leading forwards from medial iliac lymph nodes, associated with the abdominal aorta in the abdominal roof. It is also noticed in the **tracheal lymph trunks** leading back down the neck alongside the trachea from medial retropharyngeal lymph nodes associated with the pharynx.

There is, however, a tendency for paired drainage channels to join and produce a single midline vessel. Thus the lumbar lymph trunks empty into a single, enlarged, sac-like structure, the **cisterna chyli**, lying between the kidneys in the abdominal roof. This sac also receives drainage from kidneys and from guts, the latter via the large **intestinal trunk**. This structure is a derivative primarily of the jejunal lymph nodes in the mesenteric root but is often in the form of a network of vessels. The cisterna chyli is continued through the diaphragm, alongside the aorta, into the chest as the **thoracic duct**, the largest of the lymphatic channels, and continues through the thoracic roof on the left side. It eventually passes through the thoracic inlet into the base of the neck before emptying its lymph into the venous system.

The entry points for lymph into the blood system also indicate that the lymphatic system was originally paired. The **left tracheal trunk** containing lymph from the head and neck, unites with the thoracic duct before entry of the latter into the venous system at the union of the left subclavian vein and the cranial vena cava on the left side. The **right tracheal trunk** containing lymph from the right side of the head and neck, empties into the venous system with lymph from the right forelimb as the right lymphatic duct, in the angle between the right external jugular vein and the right subclavian vein.

Lymph nodes

1 Parotid lymph node – small node under parotid salivary gland (drainage area = skin and subcutaneous tissues of upper part of head, eyelids and external ear; jaw joint, jaw muscles and parotid salivary gland).

2 Mandibular lymph nodes – 2 or 3 palpable nodes up to 2 cm long at angle of jaw (drainage area = skin and subcutaneous tissues of lower part of head, lips and cheeks; mucosa of floor of mouth, tongue, jaw joint, jaw muscles and salivary glands).

3 Medial retropharyngeal lymph node – large node up to 5 cm long, deeply placed on pharynx below atlas wing (drainage area = all deep structures of head and cranial end of neck including oral, nasal and pharyngeal cavities, tongue, salivary glands, tonsils, larynx, oesophagus, orbits, deep parts of ear, muscles of head and neck. Also receives lymph from parotid and mandibular lymph nodes).

4 Superficial cervical lymph nodes – 2 nodes up to 3 cm long, cranial to shoulder and palpable if enlarged (drainage area = skin and subcutaneous tissues of caudal part of head and ear, of neck, shoulder, back and forelimb; deep structures of forelimb, shoulder and cranial part of ventral thoracic wall).

5 Deep cervical lymph nodes – cranial, middle and caudal, but small and variable in number – caudal deep cervical is only one constantly

present on trachea cranial to thoracic inlet (drainage area = deep structures of neck including bones, muscles, larynx, trachea, thyroid gland and oesophagus; deep drainage from shoulder and upper arm into caudal member of series).

6 Cranial mediastinal lymph nodes – may be as many as 6 and up to 3 cm long in cranial mediastinum on trachea (drainage area = bones and muscles of dorsal parts of neck, thorax and abdomen; trachea, oesophagus, thymus gland, heart, pericardium, aorta and parietal pleura dorsally. Also receives lymph from sternal, intercostal and tracheobronchial lymph nodes).

7 Sternal lymph node – up to 2 cm long and inside chest at lower end of intercostal space 2 (drainage area = ventral parts of thoracic wall including ribs, sternum, intercostal muscles, thoracic muscles, diaphragm and parietal pleura; also from thymus gland and pericardium. NB. May participate with axillary node in draining thoracic and cranial abdominal mammary glands. It also receives lymph from parietal peritoneum of abdomen which passes in lymphatic vessels penetrating periphery of diaphragm to run forwards beneath transverse thoracic muscles).

8 Intercostal lymph node – small and often absent, inside chest at upper end of intercostal space 5 or 6 (drainage area = bones and muscles of thoracic roof, upper parts of ribs, thoracic muscles and parietal pleura).

9 Tracheobronchial lymph nodes, left right and middle – large, in relation to tracheal bifurcation and principal bronchi (drainage area = lungs, bronchi, trachea, oesophagus, heart, aorta, mediastinum and diaphragm. Also receives lymph from pulmonary lymph nodes when present).

10 Pulmonary lymph nodes – small, inconstant at lung hilus (drainage area = lungs and bronchi).

11 Axillary lymph node – in axilla related to rib 2, palpable when enlarged (drainage area = skin and subcutaneous tissues of thoracic wall including thoracic and cranial abdominal mammary glands; deep structures of forelimb and neck. Also receives lymph from accessory axillary lymph node when present).

12 Accessory axillary lymph node – caudal to axillary in region of rib 3, inconstant (drainage area = same as for axillary).

13 Hepatic lymph nodes, left and right – flanking hepatic portal vein close to porta (drainage area = lesser curvature of stomach, duodenum, pancreas and liver).

14 Splenic lymph nodes – several on splenic vessels at hilus (drainage area = oesophagus, fundus, body and greater curvature of stomach, pancreas, liver, spleen, greater omentum and diaphragm).

15 Gastric lymph nodes – in lesser omentum close to pylorus, inconstant (drainage area = oesophagus, stomach, liver and diaphragm).

16 Pancreaticoduodenal lymph node – on cranial duodenal flexure related to pancreas, inconstant (drainage area = pancreas, duodenum and greater omentum).

17 Jejunal lymph nodes – 2 large nodes may be up to 20 cm long in a large dog in root of mesentery (drainage area = ileum, jejunum and pancreas).

18 Colic lymph nodes – several small nodes in mesocolon (drainage area = ileum, caecum, ascending and transverse colon).

19 Caudal mesenteric lymph nodes – several small nodes on colon close to pelvic inlet (drainage area = descending colon).

20 Lumbar aortic lymph nodes – 12 or more small nodes in sublumbar fat flanking aorta (drainage area = bones and muscles of abdominal roof including parietal peritoneum; ureters and abdominal parts of urogenital system including ovary, uterus and testis).

21 Renal lymph node – cranialmost member

of lumbar aortic series near renal vessels (drainage area = kidney and adrenal gland).

22 Medial iliac lymph node – up to 4 cm long in abdominal roof at pelvic inlet (drainage area = bones and muscles of dorsal half of abdomen and pelvis and thigh; abdominal and pelvic genitalia including uterus, vagina and prostate gland; caudal parts of digestive and urinary systems including urethra, bladder, colon and rectum. Also receives lymph from superficial inguinal, hypogastric and popliteal nodes).

23 Hypogastric lymph node – small in fat below lumbosacral junction in pelvic roof (drainage area = bones and muscles of pelvic roof and wall, part of loins, tail and thigh; pelvic viscera including caudal parts of urogenital and digestive tracts, vagina and vestibule, urethra, and also epididymis and vas deferens. Receives lymph from sacral nodes when present).

24 Superficial inguinal lymph nodes – usually 2, in fat in fold of groin related to vaginal process in dog and inguinal mammary gland in bitch, normally palpable (drainage area = skin and subcutaneous tissues of ventral half of abdominal wall, including caudal abdominal and inguinal mammary glands, penis, prepuce and scrotum; pelvic floor and perineum, tail, medial thigh, stifle and crus).

25 Popliteal lymph node – up to 2 cm long in popliteal fossa behind stifle joint, normally palpable (drainage area = all parts of hindlimb below stifle joint).

Major lymph trunks

26 Lumbar lymph trunk. **27** Cisterna chyli. **28** Intestinal lymph trunk. **29** Tracheal lymph trunk. **30** Right lymphatic duct. **31** Thoracic duct.

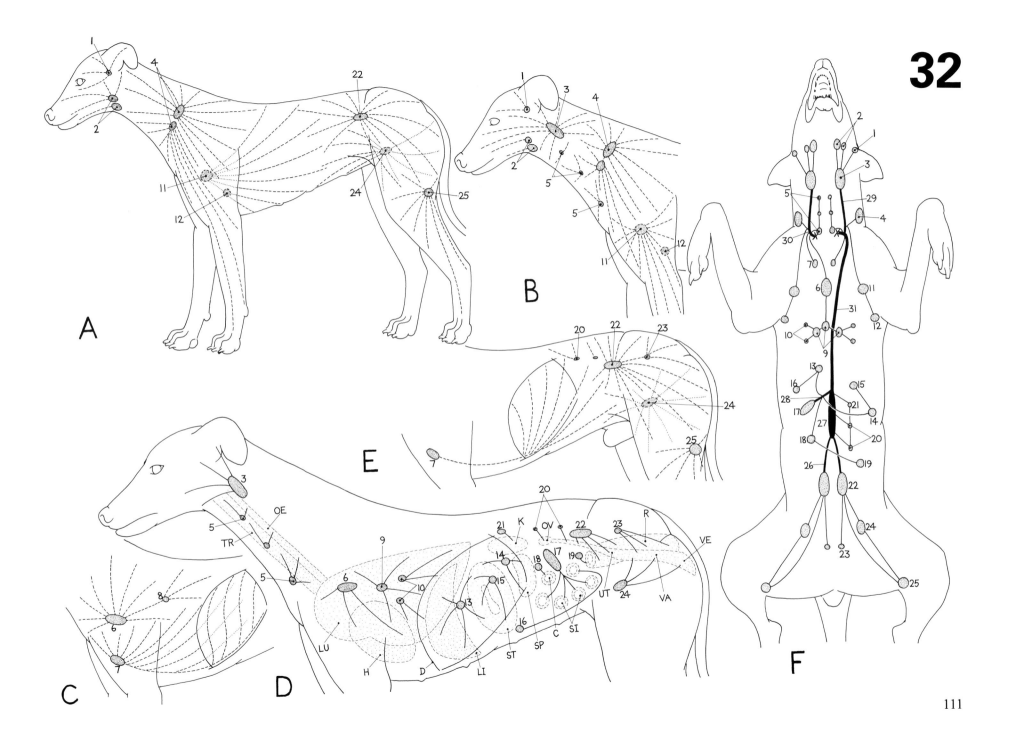

32

A

B

C

D

E

F

OE
TR

LU
H
D
LI
ST
SP
C
SI
UT
VA
VE
R
K
OV

111

33

NERVES OF THE DOG – SOMATIC DISTRIBUTION

This schema shows the major nerves of the somatic component of the peripheral nervous system. The term somatic refers to the body of the dog in distinction to the visceral parts of the body such as the respiratory system and guts. Obviously not all nerves and their branches are shown, particularly not the extensive branches from the skin. However, cutaneous nerves were shown in some detail in the drawing of the subcutaneous tissues (fig 8). Details of the brain and spinal cord (the central nervous system) and the component of the nervous system supplying viscera (the autonomic nervous system) are also not included since this would have rendered the diagram too complex by far.

The important functional units of nervous tissue are **nerve cells (neurons)** of which there are many millions throughout the body. A nerve cell has a definite cell body with a number of projections **(nerve cell fibres)** leading to and from it. These fibres are often of some considerable length and serve to link up nerve cells enabling an impulse to be passed from cell to cell. Nerve cells like any other cell type need support and binding together, by connective tissue. However, in the brain and spinal cord this connective tissue component **(neuroglia)** differs from normal connective tissue in that it consists solely of cells, extracellular material such as fibrous or elastic elements is absent. The absence of fibrous extracellular connective tissue, particularly in the brain, means that central nervous tissue is unexpectedly soft. The brains that you might view in a glass jar in a museum have hardened after death and with preservation. Organs containing fibrous connective tissue maintain their shape and form in life, as has been indicated elsewhere. Thus the nerves of the peripheral system that are grossly identifiable as they penetrate throughout the body do have a definite shape and form. Their constituent nerve fibres are bound together by normal fibrous connective tissue and surrounded by a connective tissue sheath.

What has been said above would suggest that central nervous tissue is potentially vulnerable to trauma. An obvious method of combating such a possibility has been implemented by housing the central nervous system inside the protective bony framework of the skull and vertebral column as we noticed earlier. However, to allow for vertebral movements, and to provide some measure of cushioning during such movements, the spinal cord does not fill the available space in the spinal canal. It is surrounded and enclosed by several membranes, the **meninges**, the outermost, of dense inelastic fibrous tissue, is the **dura**. This dural tube surrounding the spinal cord is anchored cranially where it merges with the periosteum of the skull bones at the foramen magnum, and caudally by merging with the periosteum of the caudal vertebrae. Where spinal nerve roots leave the cord they are surrounded by dural extensions which fuse with the periosteum of the vertebral bones at each intervertebral foramen. The dural tube is therefore stabilized and suspended in position in the spinal canal. In turn the spinal cord is anchored in position inside the dural tube by a series of bilateral ligaments attaching cord to tube.

'Cushioning' is provided for central nervous tissue in two main ways. Firstly, in the **epidural space** outside the dura, between it and the periosteum lining the vertebral canal, a quantity of fat (semifluid at body temperatures) and large longitudinal venous blood sinuses are found. (Presumably because there is no movement between skull bones an epidural space is not required in the cranium, consequently the cranial dura is fused with the periosteum of the skull bones and an epidural space is obliterated around the brain.) Secondly, inside the dural tube there is a second meningeal tube, the **arachnoid**, enclosing a **subarachnoid space** around the cord and brain which contains the **cerebrospinal fluid** forming a water cushion around central nervous tissue. The fluid maintains a pressure against the arachnoid pushing it outwards keeping it in contact with the dura and so a subdural space is normally obliterated.

Within the nervous system nerve cells have different functional responsibilities. Some nerve cells transmit sensory information from special sense organs, and from sensory receptors in the skin, peritoneum (abdominal pain), skeletal muscles and joints. Others convey motor commands to muscles and glandular (secretory) tissue. In order for a motor response to be elicited by an incoming sensation, a **sensory nerve cell** must be linked with a **motor nerve cell** either directly or through the intermediary of one or more **association nerve cells**. This broad functional division within nervous tissue is to some extent mirrored in the overall topographical arrangement of the nervous system. Thus the **peripheral nervous system** of grossly identifiable nerves is composed of a mixture of sensory and motor nerve cell fibres collected into bundles permeating throughout the body, arising from receptors and passing to muscles and glands. The **central nervous system** of brain and spinal cord is composed of incoming sensory nerve cell fibres, outgoing motor nerve cell fibres (and their cell bodies) and the entire complement of association nerve cells.

Sensory nerves actually arise from two major sources. Firstly, **'somatic'** **sensory fibres** come from receptors in or under the skin and convey impulses of touch, pressure, pain and temperature from the outside world; and from receptors within the skeletal muscles, tendons and the ligaments and capsules of joints, conveying information as to the state of tension and spatial position of the component parts. Secondly, **'visceral' sensory fibres** come from receptors in blood vessel walls, in mucous membranes, and in the substance of the internal viscera, and convey information about the state of distension or contraction of hollow viscera.

Similarly motor nerve fibres pass to two main target areas. Firstly, **'somatic'** **motor nerves** supply skeletal muscle of the body generally, a supply that is voluntary and under the conscious control or will of the dog. Secondly, **'visceral' motor nerves** supply the smooth muscle of blood vessels and viscera, the cardiac muscle of the heart, and glands throughout the body, a supply that is involuntary and not under the animal's conscious control; eg. such things as blood circulatory and digestive movements. This visceral motor supply is usually spoken of as the 'autonomic nervous system' although it is represented in both peripheral and central nervous systems, and we shall deal with it separately in fig 35.

Many of the basic activities carried out by the body and therefore controlled by the nervous system are reflex activities — automatic and not under the conscious control of the dog. This will clearly be the case if we are considering smooth muscle which by definition we have said is involuntary. Reflexes regulating visceral functions are mediated through the autonomic system with effector organs either smooth muscle, cardiac muscle or glands. Reflexes may also be of a somatic nature with voluntary muscle operating in a reflex manner. In fact much 'voluntary' muscle activity is reflex; eg. such things as the postural reflexes and reflexes involved in basic patterns of movement. A simple 'stretch reflex' is involved widely in maintaining a standing position. Weight on a forelimb for instance will tend to flex (fold) the elbow joint stretching the triceps muscle. Stretch receptors triggered in the triceps muscle send information to the spinal cord by way of the nerve to the triceps muscle (radial nerve). In the cord the impulse is transmitted to

a motor nerve cell directly and the motor impulse is conveyed to muscle fibres in the triceps muscle, again in the radial nerve, causing contraction. Many basic reflexes like this operate at a purely spinal level based on simple reflex arcs. However, more complex reflexes and the coordination and interaction between reflexes is carried out in reflex centres at higher levels in the brain — in the cerebellum for locomotory and postural reflexes.

The somatic component of the peripheral nervous system is made up of a series of paired nerves attached to the central nervous system. A **spinal nerve** actually leaves and enters the cord by two roots: a **dorsal root** dorso-laterally contains sensory fibres passing into the cord, and a **ventral root** ventro-laterally contains motor fibres leaving the cord. A dorsal root also has a thickened area where the cell bodies of the incoming sensory nerves are housed — the *dorsal root ganglion*. The cell bodies of motor nerves are actually located in the grey matter of the spinal cord. Dorsal and ventral roots combine inside the spinal canal immediately prior to their exit through intervertebral foramina between contiguous vertebrae as 'mixed' spinal nerves.

After leaving the spinal column we can see in the neck, trunk and tail, that there is a regular series of approximately 36 pairs of spinal nerves associated with the spinal cord — 8 cervical nerves, 13 thoracic nerves, 7 lumbar nerves, 3 sacral nerves, and 5 or 6 caudal nerves (the number of caudal nerves is variable). The first cervical nerve is slightly different from the rest in that it does not leave through an intervertebral foramen

but actually passes through a hole in the dorsal arch of the atlas. The second cervical nerve therefore passes through the first true intervertebral foramen between the atlas and axis, thus accounting for the presence of eight cervical nerves although only seven cervical vertebrae are present.

The spinal nerves are termed mixed spinal nerves since they contain both incoming sensory and outgoing motor fibres. Outside the spinal canal each spinal nerve branches into two major components (rami) as is shown clearly in the drawing especially in thoracic and lumbar regions. A **dorsal ramus** supplies epaxial structures — muscles, bones, fascia and skin above the level of the transverse processes: a **ventral ramus** supplies the hypaxial structures. As you will see from the schema, ventral rami are considerably more complex than dorsal, and there is a clear segmental arrangement of the distribution of the derivatives of a spinal nerve. Each nerve basically supplies sensation and motor innervation to a segment of the body corresponding to the level at which it arises from the spinal cord. Thoracic ventral rami show this segmental arrangement most clearly, providing the **intercostal nerves** running in the intercostal spaces between ribs. They supply the muscles of the chest wall and skin laterally and ventrally. The first few lumbar ventral rami continue the segmental distribution to muscles of the flank and belly and skin ventrolaterally.

Since limbs are predominantly hypaxial derivatives, ventral rami convey nerve fibres to and from them. But each limb is derived from the hypaxial

components of several body segments and so will have several ventral rami supplying it. The ventral rami to a limb tend to join up and form an interconnected plexus from which the named nerves to the limb arise:

Forelimb supplied from the **brachial plexus** — an interconnection of ventral rami from four spinal nerves C6—T1.

Hindlimb supplied from the **lumbo-sacral plexus** — an interconnection of ventral rami from seven spinal nerves L4—S3.

Aside from their contribution to the brachial plexus caudal cervical nerves also form the **phrenic nerve** which runs caudally in the mediastinum *en route* for the diaphragm which it innervates. The first two cervical nerves also spread sensory ramifications onto the pinna of the ear, the throat and the face.

Aside from their contribution to the sacral part of the lumbosacral plexus, sacral ventral rami also provide innervation to the perineum and external genitalia through a **pudendal nerve** and its perineal branches.

The ventral rami of the caudal nerves innervate the depressor muscles of the tail. Despite there being anything up to twenty or more caudal vertebral segments there are at the most five caudal spinal nerves and the ventral rami tend to join up to produce a plexus.

Innervation within the head is more complex and based on the ramifications of twelve pairs of **cranial nerves** attached to the brain. These are not as regularly arranged as spinal nerves and exit from the cranial cavity through several openings in the base of the skull. The peripheral distribution of only the largest of the cranial nerves are indicated on the diagram. Six of the cranial nerves have a very restricted distribution and are not considered in our drawing: the olfactory (cranial nerve 1), optic (cranial nerve 2) and auditory (cranial nerve 8) supply the nose, eye and ear respectively, the organs of special sensation; the oculomotor (cranial nerve 3), trochlear (cranial nerve 4) and abducent (cranial nerve 6) supply the muscles that move the eyeball in its socket.

Of the remaining six nerves the **trigeminal nerve** (*cranial nerve 5*) has the most extensive distribution in the head. It is responsible for conveying sensation from the skin of the head and the mucosa of the mouth cavity, and motor impulses to the jaw muscles. A number of its ramifications are shown in the schema. The **facial nerve** (*cranial nerve 7*) is also an extensively distributed nerve since it is responsible for conveying motor innervation to the muscles of facial expression. Some of its ramifications are shown in the schema. The **hypoglossal nerve** (*cranial nerve 12*), also shown, innervates tongue muscles.

Certain of the cranial nerves also supply structures in the neck. The connection (*cervical loop*) between the hypoglossal nerve and cervical nerve 1 is shown in the diagram. Fibres from the loop pass to strap muscles on the underside of the throat and neck. The **accessory nerve** (*cranial nerve 11*) is also shown on the diagram extending back in the neck, communicating with the ventral rami of cervical nerves, and innervating certain muscles of the lateral surface of the neck and shoulder. The two remaining cranial nerves, glossopharyngeal (cranial nerve 9) and vagus (cranial nerve 10) are not included in the drawing. The vagus, widely distributing autonomic fibres to the viscera, is shown in fig 35B.

Dorsal rami of spinal nerves with cutaneous branches and muscular branches to epaxial muscles

1 Dorsal ramus of cervical nerve 1 (suboccipital nerve). **2** Dorsal ramus of cervical nerve 2 (greater occipital nerve). **3** Dorsal ramus of cervical nerve 7 (NB dorsal ramus of last cervical nerve [C8] always small and often absent). **4** Dorsal rami of thoracic nerves 1 and 13. **5** Dorsal rami of lumbar nerves 1 and 7 (cutaneous branches are cranial clunial nerves). **6** Dorsal rami of sacral nerves 1 and 3 (cutaneous branches are middle clunial nerves.) **7** Dorsal rami of caudal nerves 1–5 (interconnected to form a dorsal caudal plexus [trunk]).

Ventral rami of spinal nerves

8 Ventral ramus of cervical nerve 1. **9** Ventral ramus of last cervical nerve (C8). **10** Cervical loop (ansa hypoglossi – connecting ventral ramus of cervical nerve 1 with hypoglossal nerve [cranial nerve 12]). **11** Cervical plexus (interconnections between ventral rami of cervical nerves). **12** Great auricular nerve (from ventral ramus of cervical nerve 2). **13** Transverse cervical nerve (from ventral ramus of cervical nerve 2). **14** Supraclavicular nerves (from ventral rami of cervical nerves 3–5). **15** Phrenic nerve (from ventral rami of cervical nerves 5–7). **16–36** Nerves arising from brachial plexus – ventral rami of C6–T1. **16** Subclavian nerve (nerve to brachiocephalic muscle from ventral ramus of cervical nerve 6). **17** Suprascapular nerve (from ventral rami of cervical nerves 6–7). **18** Musculocutaneous nerve (from ventral rami of cervical nerves 7–8). **19** Medial cutaneous antebrachial nerve (from musculocutaneous nerve). **20** Axillary nerve (from ventral rami of cervical nerves 7–8). **21** Cranial lateral cutaneous brachial nerve (from axillary nerve). **22** Pectoral nerve (from ventral rami of cervical nerve 8 and thoracic nerve 1 and incorporated with lateral thoracic nerve at its origin). **23** Radial nerve (from ventral rami of cervical nerves 7–8 and thoracic nerve 1). **24** Deep (motor) branch of radial nerve. **25** Superficial (cutaneous) branch of radial nerve. **26** Lateral and medial components of superficial branch of radial nerve. **27** Lateral cutaneous antebrachial nerve (from radial nerve). **28** Median nerve (from ventral rami of cervical nerve 8 and thoracic nerves 1–2). **29** Ulnar nerve (from ventral rami of cervical nerve 8 and thoracic nerves 1–2). **30** Caudal cutaneous antebrachial nerve (from ulnar nerve). **31** Dorsal carpal branch of ulnar nerve. **32** Palmar branch of ulnar nerve. **33** Dorsal common digital nerves. **34** Dorsal proper digital nerves. **35** Palmar common digital nerves. **36** Palmar proper digital nerves. **37** Ventral ramus of 1st thoracic (intercostal) nerve. **38** Ventral ramus of 12th thoracic intercostal) nerve. **39** Costoabdominal nerve (ventral ramus of last thoracic nerve – T13). **40** Intercostal nerve 7 (ventral ramus of thoracic nerve 7). **41** Distal lateral cutaneous branches of intercostal and lumbar nerves. **42** Ventral cutaneous branches of intercostal nerves 3–10. **43** Cranial iliohypogastric nerve (ventral ramus of lumbar nerve 1). **44** Caudal iliohypogastric nerve (ventral ramus of lumbar nerve 2). **45** Ilioinguinal nerve (ventral ramus of lumbar nerve 3). **46** Ventral ramus of last lumbar nerve (L7). **47** Genitofemoral nerve (external spermatic nerve from ventral ramus of lumbar nerves 3–4). **48** Lateral cutaneous femoral nerve (from ventral ramus of lumbar nerve 4). **49** Ventral rami of sacral nerves 1–3. **50–69** Nerves arising from lumbosacral plexus – ventral rami of L3–S3. **50** Femoral nerve (from ventral rami of lumbar nerves 5–6). **51** Saphenous nerve (from femoral nerve). **52** Obturator nerve (from ventral rami of lumbar nerves 5–6). **53** Lumbosacral trunk (from ventral rami of lumbar nerves 6–7 and sacral nerves 1–2). **54** Cranial gluteal nerve (from lumbosacral trunk). **55** Caudal gluteal nerve (from lumbosacral trunk). **56** Ischiatic nerve (direct continuation outside pelvis of lumbosacral trunk). **57** Motor branches of ischiatic nerve. **58** Common peroneal nerve (fibular nerve from ischiatic nerve). **59** Lateral cutaneous sural nerve (from common peroneal nerve or from ischiatic nerve). **60** Superficial peroneal (fibular) nerve (from common peroneal nerve). **61** Deep peroneal (fibular) nerve (from common peroneal nerve). **62** Tibial nerve (from ischiatic nerve). **63** Caudal cutaneous sural nerve (from tibial nerve or from ischiatic nerve). **64** Plantar nerves (medial and lateral from tibial nerve). **65** Dorsal common digital nerves. **66** Dorsal proper digital nerves. **67** Plantar common digital nerves. **68** Plantar metatarsal nerves. **69** Plantar proper digital nerves. **70** Autonomic branch from ventral ramus of sacral nerve to pelvic plexus. **71** Pudendal nerve (from ventral rami of sacral nerves 1–3). **72** Caudal cutaneous femoral nerve (from ventral rami of sacral nerves 1–2 giving rise to caudal clunial nerves). **73** Ventral rami of caudal nerves 1–5 (interconnected to form a ventral caudal plexus or trunk).

Cranial nerves

74–76 Trigeminal nerve (cranial nerve 5). **74** Ophthalmic component of trigeminal nerve. **75** Maxillary component of trigeminal nerve. **76** Mandibular component of trigeminal nerve. **77** Facial nerve (cranial nerve 7). **78** Spinal accessory nerve (cranial nerve 11). **79** Hypoglossal nerve (cranial nerve 12).

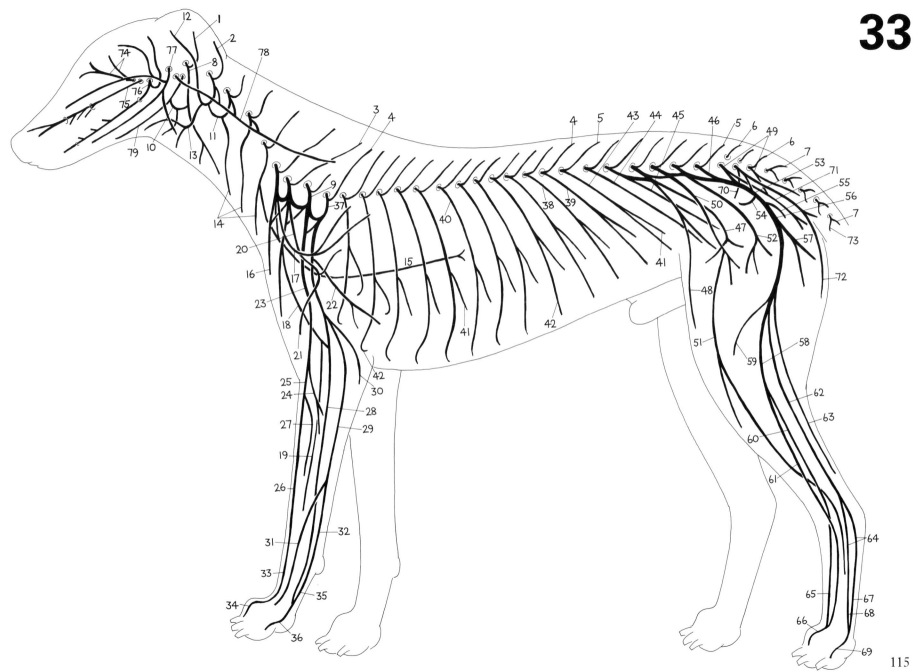

33

115

34

MOTOR INNERVATION OF THE SKELETAL MUSCLES OF THE DOG

The groups of body muscles innervated by specific nerves are indicated on these drawings in various different types of shading. If you remember when muscles were considered earlier the emphasis was firmly placed on groups of muscles acting together to produce actions at specific joints. The drawing here through the medium of its shading, shows how the motor nerve innervation to muscles is organized in terms of these same muscle groups. In practical terms what this means is that should damage occur to a particular nerve then specific groups of muscles will be affected. Temporary or even permanent paralysis of muscles (loss of voluntary movement) might well be a symptom of such nerve damage.

Muscle groups produce specific actions at particular joints so that disruption of nerve supply and the consequent muscle paralysis will produce alterations in the action of these joints. Certain actions may be lost by a joint; other actions may be accentuated at that joint because of the loss of antagonistic muscle activity.

Should you examine the drawing, and in particular the forelimb, you might notice that the radial nerve is very important because of the muscle groups that it innervates; viz. the extensor muscles of the elbow (triceps group), carpal and digital joints. A paralysis involving the radial nerve leads to a dog being unable to perform the actions of elbow, carpal and digital extension. Loss of the capacity to extend its carpal and digital joints will accentuate the action of the opposing flexors of these joints, and a dog so afflicted might well stand and move 'knuckled over'; ie. placing and dragging the dorsum of its paw along the ground. Loss of the capacity to extend its elbow, on the other hand, is a far more serious situation for a dog to cope with since it means that the elbow joint cannot be fixed in position (stabilized) to support weight. Weight placed on the limb will automatically flex the elbow and collapse the limb. The dog would limp along very badly on three legs only.

Muscles innervated by cranial nerves
1–3 Muscles innervated by trigeminal nerve (cranial nerve 5). 1 Masseter muscle. 2 Temporal muscle. 3 Digastric muscle. 4–7 Muscles innervated by facial nerve (cranial nerve 7). 4 Levator muscle of upper lip. 5 Orbicularis oris muscle. 6 Buccinator muscle. 7 Orbicularis oculi muscle. 8–9 Muscles innervated by vagus nerve (cranial nerve 10). 8 Pharyngeal muscles. 9 Cricothyroid muscle. 10–13 Muscles innervated by spinal accessory nerve (cranial nerve 11). 10 Sternocephalic muscle. 11 Brachiocephalic muscle (cleidocervical part). 12 Omotransverse muscle. 13 Trapezius muscle (cervical and thoracic parts). 14 Geniohyoid muscle innervated by hypoglossal nerve (cranial nerve 12).

Muscles innervated by nerves arising from brachial plexus
15 Ventral serrate muscle (thoracic part) innervated by long thoracic nerve. 16 Latissimus dorsi muscle innervated by dorsal thoracic nerve. 17–18 Muscles innervated by pectoral nerves. 17 Superficial pectoral muscle. 18 Deep pectoral muscle. 19–20 Muscles innervated by suprascapular nerve. 19 Supraspinatus muscle. 20 Infraspinatus muscle. 21–24 Muscles innervated by axillary nerve. 21 Deltoid muscle. 22 Teres major muscle. 23 Teres minor muscle. 24 Brachiocephalic muscle (cleidobrachial part). 25–26 Muscles innervated by musculocutaneous nerve. 25 Biceps brachii muscle. 26 Brachial muscle. 27–32 Muscles innervated by radial nerve. 27 Triceps brachii muscle. 28 Radial carpal extensor muscle. 29 Common and lateral digital extensor muscles. 30 Ulnar carpal extensor muscle (lateral ulnar). 31 Oblique carpal extensor muscle. 32 Supinator muscle. 33–36 Muscles innervated by median nerve. 33 Pronator teres muscle. 34 Radial carpal flexor muscle. 35 Superficial digital flexor muscle. 36 Deep digital flexor muscle (all three heads). 37–38 Muscles innervated by ulnar nerve. 37 Ulnar carpal flexor muscle. 38 Deep digital flexor muscle (bar radial head).

Muscles innervated by dorsal rami of spinal nerves
39 Splenius muscle. 40 Iliocostal and longissimus muscles. 41 Dorsal sacrocaudal muscles.

Muscles innervated by ventral rami of spinal nerves
42 Sternohyoid muscle. 43 Sternothyroid muscle. 44 Thyrohyoid muscle. 45 Longus colli muscle. 46 Scalene muscle. 47 Ventral serrate muscle (cervical part). 48 Rhomboid muscle (capital, cervical and thoracic parts). 49 Dorsal serrate muscle. 50 External intercostal muscles. 51 External abdominal oblique muscle. 52 Internal abdominal oblique muscle. 53 Rectus abdominis muscle. 54 Coccygeus muscle. 55 Ventral sacrocaudal muscles.

Muscles innervated by nerves arising from lumbosacral plexus
56–59 Muscles innervated by cranial gluteal nerve. 56 Middle gluteal muscle. 57 Deep gluteal muscle. 58 Piriform muscle. 59 Tensor muscle of lateral femoral fascia. 60 Superficial gluteal muscle innervated by caudal gluteal nerve. 61–63 Muscles innervated by femoral nerve. 61 Iliopsoas muscle. 62 Quadriceps femoris muscle. 63 Sartorius muscle. 64 Adductor muscles innervated by obturator nerve. 65–67 Muscles innervated by rotator nerves from ischiatic nerve. 65 Quadratus femoris muscle. 66 Internal obturator muscle. 67 Gemelli muscles. 68–70 Muscles innervated by ischiatic nerve. 68 Biceps femoris muscle. 69 Semitendinosus muscle. 70 Semimembranosus muscle. 71–73 Muscles innervated by tibial nerve. 71 Gastrocnemius muscle. 72 Superficial digital flexor muscle. 73 Deep digital flexor muscle. 74–76 Muscles innervated by peroneal (fibular) nerve. 74 Cranial tibial muscle. 75 Long and lateral digital extensor muscles. 76 Long peroneal (fibular) muscle.

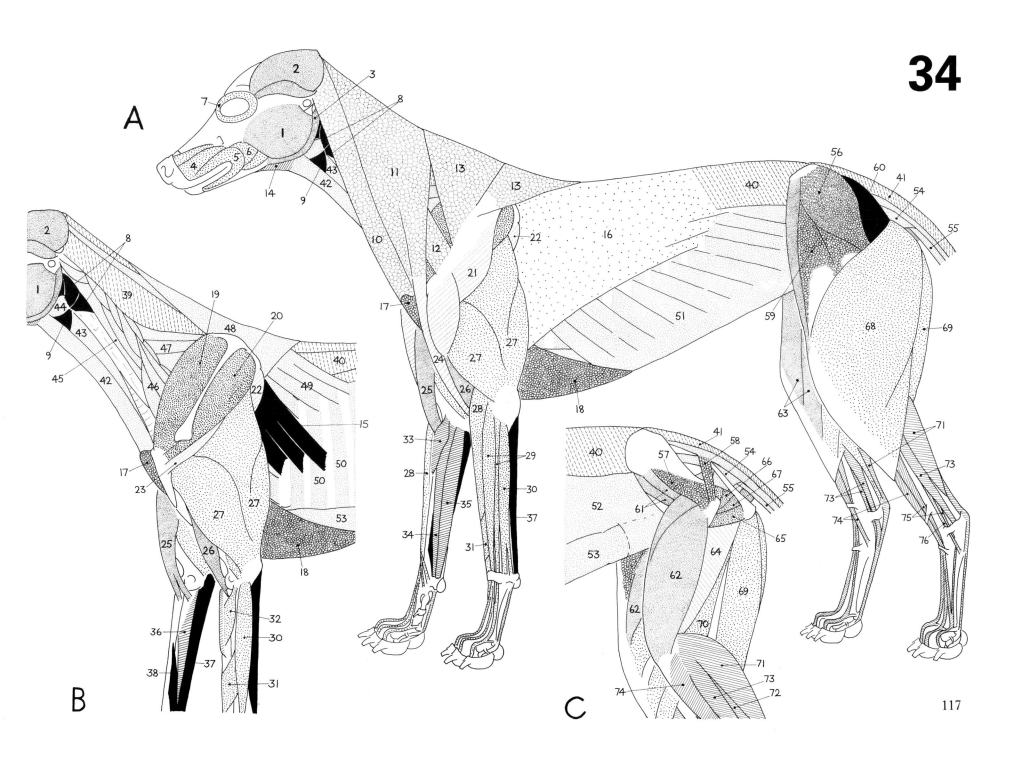

35

NERVES OF THE DOG – VISCERAL (AUTONOMIC) DISTRIBUTION

The autonomic is that component of the nervous system concerned with maintaining a constant internal environment. In order to do this it exerts motor control over smooth muscle, cardiac muscle and glands throughout the body. Structures supplied will include muscles in blood vessel walls throughout the body (including the vessels in the skin and in skeletal muscles). In the skin the smooth muscles associated with hairs, and the sweat and sebaceous glands are also innervated. Strictly 'visceral' structures receiving autonomic innervation will include the 'normal' smooth muscles associated with all parts of the respiratory, digestive and urogenital systems, in addition to specialized areas of smooth muscle such as the pupillary and ciliary muscles in the eyeball, cardiac musculature in the heart, and sphincter muscles wherever they occur. Other visceral structures innervated include glands such as the lacrimal and salivary in the head, bronchial in the thorax, liver and pancreas in the abdomen, and prostate in the pelvis. The muscles in the walls of the blood vessels to all of these viscera are also innervated.

This visceral motor supply is organized on both functional and anatomical grounds as two systems – a **sympathetic system** and a **parasympathetic system**. Functionally the two systems tend to have opposing actions so that when an organ or tissue receives both types of innervation the effects are usually antagonistic; eg. the heart is inhibited (slowed) by parasympathetic innervation and excited (speeded up) by sympathetic innervation. Anatomically the two systems differ in the regions of the central nervous system from which they originate.

Sympathetic system – nerve fibres arising from the spinal cord in the thoracic and lumbar regions back to and including lumbar segment 4 (*thoracolumbar autonomic outflow*).
Parasympathetic system – nerve fibres arising from the brain and from the spinal cord in the sacral region (*craniosacral autonomic outflow*).

Anatomical differences are also apparent in the overall distribution of the two components. Sympathetic fibres are widely distributed to the whole body both somatic and visceral, whereas parasympathetic fibres are only distributed to the viscera, they do not supply skin or body musculature.

The feature that distinguishes the autonomic from the peripheral somatic nervous system dealt with earlier, is that autonomic pathways have a relay in them – a **synapse**. These relays are collected into specific **peripheral ganglia**, which are often grossly identifiable, and so we can recognize **preganglionic nerve fibres** passing from the central nervous system to a ganglion, and **postganglionic nerve fibres** extending from a ganglion to a target organ. It is of interest to note that in the sympathetic system for every preganglionic fibre there are approximately twenty postganglionic fibres, indicative of the widespread nature of sympathetic activity. This contrasts with the specifically localized activity of the parasympathetic system where the ratio is as little as two postganglionics for every preganglionic fibre. The restricted activity of the parasympathetic system is also evident in the location of its synaptic relays in **terminal ganglia** either on or in the substance of the target organ. Consequently these are the peripheral ganglia we are unable to identify grossly unlike the sympathetic ganglia which are usually identifiable and not in the substance of their target organs.

If we consider initially the **sympathetic system** shown in the upper drawing (**A**), sympathetic nerve fibres leaving the spinal cord as the thoracolumbar outflow are channelled into the ventral rami of the corresponding spinal nerves (T1-L4). These sympathetic fibres leave a ventral ramus through a **visceral (communicating) ramus** extending ventrally around the vertebral body to enter an expanded **vertebral ganglion** ventrolateral to the vertebral body. Individual sympathetic fibres are surrounded by a fatty sheath and because of this the visceral rami containing them are termed '*white rami*'. Sympathetic fibres need not synapse in this ganglion of entry but may pass either cranially or caudally to ganglia further forwards or backwards in the body. These fibres collected together link the vertebral ganglia into a **sympathetic trunk**. The paired trunks with periodic ganglionic enlargements are the most conspicuous parts of the system.

Somatic distribution from sympathetic trunk ganglia (to smooth muscle in blood vessels and skin) involves reentry of postganglionic fibres into the visceral rami and so back into spinal nerves for subsequent distribution with their dorsal and ventral branches. In the drawing postganglionic nerve distribution is shown by the broken lines and in the thoracolumbar region (back to and including lumbar segment 4) postganglionic fibres accompany preganglionic fibres in the white visceral rami. However preganglionic fibres only leave the spinal cord between T1 and L4, whereas postganglionics are distributed to blood vessels and skin throughout the body, in head, neck, trunk and tail. Therefore the *sympathetic trunks extend the whole length of the body with segmental vertebral ganglia linked by visceral rami to all of the spinal nerves.* Visceral rami outside the thoracolumbar region only contain postganglionic fibres which lack fatty insulating sheaths around them and are referred to as *grey rami*.

As you can see from the drawing, in certain parts of the sympathetic trunk the orderly arrangement of ganglia is lost. This is especially apparent in the neck where some combining has occurred to give two enlarged ganglionic masses: a **cranial cervical sympathetic ganglion** at the cranial end of the neck close to the base of the skull, and a **middle cervical sympathetic ganglion** at the base of the neck close to the thoracic inlet. Between the two the sympathetic trunk runs in the carotid sheath with the vagus nerve as a component of the **vagosympathetic trunk**.

In the thorax a caudal cervical and several cranial thoracic sympathetic ganglia coalesce to produce an enlarged **cervicothoracic (stellate) ganglion** at the upper end of the first intercostal space. Distribution of sympathetic fibres to skin and blood vessels of the neck originates primarily from the cervicothoracic ganglion via the **vertebral nerve** which is merely a collection of postganglionic (grey) visceral rami. The vertebral nerve accompanies the vertebral artery along the neck in the transverse canal of the neck vertebrae. At each intervertebral space it links with a cervical spinal nerve for peripheral distribution.

The remaining thoracic part of the sympathetic trunk has a regular segmental arrangement of ganglia and the trunk passes back into the abdominal roof. Unlike the aorta, oesophagus and caudal vena cava, which penetrate through the diaphragm the sympathetic trunk passes dorsal to it. The lumbar part of the trunk also has fairly regularly arranged ganglia but in the sacral and caudal regions the trunks themselves as well as the ganglia become somewhat irregular in their arrangement. Just as in the neck, the visceral rami in the caudal lumbar and sacral regions are 'grey rami' only carrying postganglionic fibres to the spinal nerves.

Sympathetic distribution to viscera within the body cavities is in **visceral nerves** leaving the sympathetic trunks. In the thorax visceral nerves to the heart and lungs originate from the middle cervical, cervicothoracic and thoracic ganglia and contain postganglionic fibres. These produce cardiac and pulmonary plexuses of nerves on the base of the heart and on the bronchi at the roots of the lungs. In the abdomen preganglionic fibres do not relay in the sympathetic trunk ganglia but leave in **splanchnic nerves** and pass to **prevertebral (collateral) ganglia** associated with the aorta and the major arteries leaving it especially the three main visceral arteries – *coeliac, cranial mesenteric and caudal mesenteric ganglia*. The splanchnic nerves are variable but generally originate from the more caudal thoracic ganglia and also from cranial lumbar ganglia. Postganglionic fibres are distributed from the prevertebral ganglia to the abdominal viscera and their blood vessels as plexuses in or on the walls of the blood vessels themselves. In this way they follow the branches of the arteries to the target organs.

In order to reach viscera in the pelvis preganglionic fibres pass through the caudal mesenteric ganglia and enter a pair of **hypogastric nerves** which travel back in the mesentery of the descending colon. The hypogastric nerves terminate on either side of the rectum in a collection of ganglia the **pelvic plexus** on the lateral rectal wall. Postganglionic distribution from this plexus to pelvic viscera including the external genitalia is primarily with branches of the internal iliac artery.

In the head and cranial end of the neck the viscera and vessels receive postganglionic sympathetic fibres from the cranial cervical sympathetic ganglion primarily distributed as plexuses on or in the walls of branches of the internal and external carotid arteries. Some postganglionic fibres will link with cranial nerves emerging from the skull in the neighbourhood of the ganglion itself to be distributed with their branches. In drawing **A**, sympathetic fibres are shown passing through parasympathetic ganglia without a relay having already synapsed in the cranial cervical sympathetic ganglion.

The lower drawing (**B**) shows the **parasympathetic system** and you will see that parasympathetic fibres leaving the brain are distributed in four of the twelve cranial nerves (oculomotor [3], facial [7], glossopharyngeal [9], and vagus [10]). This distribution as you can see, not only supplies viscera in the head but also in the neck, thorax and abdomen. In this respect the **vagus nerve** is the most important, passing down the neck in company with the sympathetic trunk as the vagosympathetic trunk we have already mentioned. At the thoracic inlet the vagus leaves the sympathetic trunk and passes back in the mediastinum with the oesophagus on which it divides into dorsal and ventral vagal trunks before entering the abdomen. *En route* through thorax and abdomen the vagus nerve gives rise to numerous branches supplying viscera as far caudally as the descending colon. Vagal distribution takes place in association with postganglionic sympathetic fibres in the plexuses in blood vessel walls.

In the thorax vagal fibres are distributed with sympathetic fibres in the cardiac and pulmonary plexuses. In the abdomen the dorsal and ventral vagal trunks are distributed with the plexuses of the main blood vessels. Vagal fibres pass through the prevertebral sympathetic ganglia without synapsing since parasympathetic relays normally occur in terminal ganglia located on or in the viscus supplied. These terminal ganglia are therefore not normally grossly identifiable and postganglionic fibres are short. The four peripheral parasympathetic ganglia found in the head are exceptions to this since they are of some size and at some remove from their targets. A **ciliary ganglion** relaying to the smooth muscle in the eyeball: **pterygopalatine ganglion** relaying fibres to the lacrimal and nasal glands: **submandibular ganglion** relaying fibres to the mandibular and sublingual salivary glands: **otic ganglion** relaying fibres to the zygomatic and parotid salivary glands.

Parasympathetic fibres leaving the spinal cord in the sacral region travel in the ventral rami of the three sacral spinal nerves to leave in pelvic nerves which terminate in the ganglionic masses of the pelvic plexus. Here they intermingle with sympathetic fibres from the hypogastric nerves and with them postganglionic fibres are distributed to the pelvic viscera along branches of the internal iliac artery.

Although strictly the autonomic system applies to motor pathways, the whole system functions on the anatomical basis of reflex arcs like any other part of the nervous system. Visceral functions are controlled by sensory information in just the same manner as somatic functions, and visceral sensory fibres accompany sympathetic and parasympathetic nerve fibres. Such sensory information will arise from both physical stimuli such as heat and pressure, as well as chemical stimuli such as carbon dioxide or oxygen concentrations. Autonomic functions under reflex control include such things as respiration, blood pressure regulation, gut activity, glandular secretion, bladder emptying, and so on.

Although not included in the drawing the **adrenal glands** are of importance in any consideration of the autonomic nervous system. They are shown in fig 27E in the abdominal roof between the kidneys, related to the aorta and caudal vena cava. Their importance lies in the role they play in reinforcing the effects of sympathetic activity. The central core of an adrenal gland, the **adrenal medulla**, is formed in embryonic development from the same cellular rudiments that produce the peripheral sympathetic ganglia and postganglionic sympathetic nerves. What this means is that the adrenal medulla will function as a giant postganglionic sympathetic fibre passing adrenal hormone (*adrenaline* and *noradrenaline*) into the bloodstream for widespread distribution. The effects produced are superimposed on the more local effects produced by sympathetic postganglionic nerves. You may surmise from this that the adrenal medulla must also receive preganglionic fibres from the sympathetic trunk in visceral nerves. These actually originate from the more caudal thoracic segments of the sympathetic trunk as part of the splanchnic nerves.

Ventral rami of spinal nerves
1 Cervical nerve 1. **2** Cervical nerve 8. **3** Thoracic nerve 1. **4** Lumbar nerve 4. **5** Sacral nerve 1.

Sympathetic nervous system
6–11 Visceral (communicating) rami of spinal nerves. **6–9** Grey visceral rami (postganglionic sympathetic fibres only). **6** Visceral ramus of cervical nerve 1. **7** Visceral ramus of cervical nerve 8. **8** Visceral ramus of lumbar nerve 5. **9** Transverse (vertebral) nerve (combined grey visceral rami of cervical nerves 3–7). **10–11** White visceral rami (preganglionic and postganglionic sympathetic fibres intermingled). **10** Visceral ramus of thoracic nerve 1 (1st white visceral ramus). **11** Visceral ramus of lumbar nerve 4 (last white visceral ramus). **12–14** Sympathetic trunk. **12** Sympathetic trunk in thorax, abdomen and pelvis (trunk in pelvis becomes irregular and trunks of left and right sides may join). **13** Sympathetic trunk in neck (runs with vagus nerve as vagosympathetic trunk). **14** Subclavian loop (subdivision of sympathetic trunk around subclavian artery). **15–20** Sympathetic trunk ganglia. **15** Cranial cervical sympathetic ganglion. **16** Middle cervical sympathetic ganglion. **17** Cervicothoracic (stellate) sympathetic ganglion (fused ganglia of C8-T3). **18** Thoracic sympathetic ganglion. **19** Lumbar sympathetic ganglion. **20** Sacral sympathetic ganglion (ganglia in pelvis become irregular and may combine). **21–24** Splanchnic nerves. **21** Greater splanchnic nerve. **22** Lesser splanchnic nerve. **23** Lumbar splanchnic nerves. **24** Sacral splanchnic nerves. **25–30** Collateral sympathetic ganglia and plexuses. **25–26** Coeliacomesenteric plexus. **25** Coeliac ganglion. **26** Cranial mesenteric ganglion. **27** Caudal mesenteric ganglion. **28** Pelvic plexus. **29** Intermesenteric (abdominal aortic) plexus. **30** Hypogastric nerve. **31–37** Postganglionic sympathetic distribution to viscera. **31** Postganglionic sympathetic distribution to viscera in head and cranial end of neck with carotid artery. **32–33** Postganglionic sympathetic distribution to viscera in neck and thorax. **32** Cardiosympathetic nerves. **33** Bronchosympathetic nerves. **34–36** Postganglionic sympathetic distribution to abdominal viscera. **34** Sympathetic distribution with branches of coeliac artery. **35** Sympathetic distribution with branches of cranial mesenteric artery. **36** Sympathetic distribution with branches of caudal mesenteric artery. **37** Postganglionic sympathetic distribution to pelvic viscera with branches of urogenital artery.

Parasympathetic nervous system
38–52 Cranial parasympathetic outflow to glandular structures in head. **38** Preganglionic parasympathetic fibres in oculomotor nerve (3). **39** Ciliary ganglion. **40** Postganglionic parasympathetic fibres to ciliary and pupillary muscles in short ciliary and nasociliary nerves. **41** Preganglionic parasympathetic fibres originating in facial nerve (7) but distributed in branches of maxillary component of trigeminal nerve (5) – vectored from facial through nerve of pterygoid canal. **42** Pterygopalatine ganglion. **43** Postganglionic parasympathetic fibres to nasal glands in caudal nasal nerve. **44** Postganglionic parasympathetic fibres to palatine glands in major and minor palatine nerves. **45** Postganglionic parasympathetic fibres to lacrimal gland in lacrimal nerve. **46** Preganglionic parasympathetic fibres originating in facial nerve (7) and distributed in mandibular component of trigeminal nerve (5) – vectored from facial through chorda tympani nerve and lingual nerve. **47** Mandibular ganglion. **48** Postganglionic parasympathetic fibres to mandibular and sublingual salivary glands in lingual nerve. **49** Preganglionic parasympathetic fibres originating in glossopharyngeal nerve (9) but distributed in branches of trigeminal nerve (5). **50** Otic ganglion. **51** Postganglionic parasympathetic fibres to parotid salivary gland in auriculotemporal nerve. **52** Postganglionic parasympathetic fibres to zygomatic salivary gland in buccal nerve. **53–64** Cranial parasympathetic outflow distributing to viscera of neck, thorax and abdomen caudally as far as transverse colon. **53** Vagus nerve (10). **54** Cranial laryngeal nerve. **55** Vagus nerve in neck (runs with cervical sympathetic trunk as vagosympathetic trunk). **56** Recurrent laryngeal nerve. **57** Tracheal and oesophageal branches from recurrent laryngeal nerve. **58** Caudal laryngeal nerve (termination of recurrent laryngeal nerve). **59** Cardiovagal nerve. **60** Bronchial and pulmonary branches of vagal nerve. **61** Dorsal vagal trunk (trunks of left and right sides join close to diaphragm). **62** Ventral vagal trunk (trunks of left and right sides join close to diaphragm). **63–64** Preganglionic parasympathetic vagal distribution to abdominal viscera accompanying postganglionic sympathetic fibres. **63** Parasympathetic distribution with branches of coeliac artery. **64** Parasympathetic distribution with branches of cranial mesenteric artery. **65–67** Sacral parasympathetic outflow distributing to pelvic and caudal abdominal viscera. **65** Pelvic splanchnic nerves from sacral spinal nerves 1–3. **66–67** Parasympathetic distribution from pelvic plexus accompanying postganglionic sympathetic fibres. **66** Parasympathetic distribution with branches of urogenital artery. **67** Parasympathetic distribution through hypogastric nerves, caudal mesenteric ganglion and with branches of caudal mesenteric artery. **68** Left subclavian artery. **69** Heart. **70** Aorta. **71** Larynx. **72** Trachea. **73** Lung. **74** Diaphragm. **75** Oesophagus. **76** Stomach. **77** Duodenum. **78** Jejunum/Ileum. **79** Colon. **80** Rectum. **81** Anal canal. **82** Liver. **83** Spleen. **84** Kidney. **85** Bladder. **86** Urethra. **87** Ovary. **88** Uterus. **89** Vestibule. **90** Mandibular salivary gland. **91** Sublingual salivary gland. **92** Parotid salivary gland. **93** Zygomatic salivary gland. **94** Nasal glands. **95** Palatine glands. **96** Lacrimal gland.

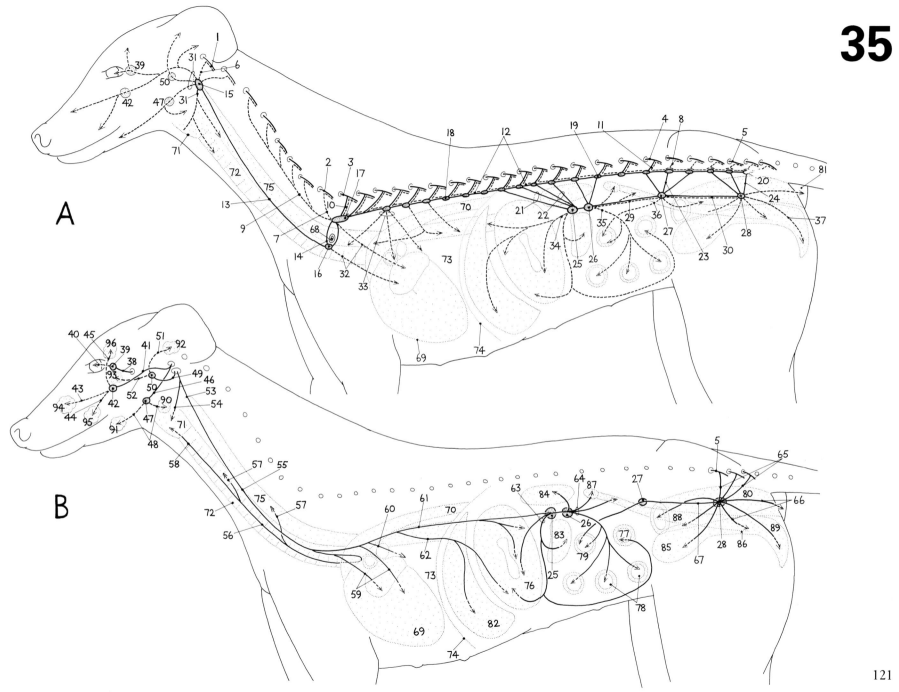

36

SURFACE ANATOMY OF THE DOG – HEAD, NECK AND TRUNK

This drawing, and the final two (figs 37 & 38), attempt to portray most of the structures that are clearly related to the body surface, either palpable or visualizable in surface projection.

Surface features
1 Nasal plane (pigmented hairless skin). 2 External nostril (leading into nasal vestibule surrounded by nasal cartilages). 3 Philtrum. 4 Pinna (visible part of external ear). 5 External opening of ear canal. 6 Jugular fossa (triangular depression at base of neck). 7 Axilla. 8 Median pectoral groove. 9 Umbilicus (navel – hairless scar denoting point of entry and exit of blood vessels in foetus). 10 Fold of flank. 11 Fold of groin. 12 Position of superficial (subcutaneous) opening of inguinal canal. 13 Femoral triangle (bordered by sartorius, pectineus and abdominal wall muscles).

Bones, joints and ligaments of head
14 Nasal cartilages (movably articulated with bone of nasal cavity and surrounding nasal vestibule). 15 Nasal process of incisive bone (bordering bony nasal opening leading into nasal cavity proper). 16 Infraorbital foramen (passage of infraorbital branches of maxillary artery and nerve). 17 Bony orbital margin. 18 Zygomatic (supraorbital) process of frontal bone. 19 Zygomatic arch (bar of bone connecting face and cranium below eye). 20 Orbital ligament (joining frontal bone and zygomatic arch and completing orbital rim). 21 Temporal line (rostral divergence of external sagittal crest). 22 External sagittal crest (in dorsal midline of cranium). 23 External occipital protuberance (occiput). 24 Nuchal crest (division between dorsal and caudal surfaces of cranium). 25 Mastoid process of temporal bone (sole representation on skull surface of petrous temporal bone). 26–28 Mandible (lower jaw). 26 Mandibular body. 27 Angular process of mandible. 28 Mandibular symphysis. 29 Mental foramen (passage of mental branches of mandibular alveolar nerve and vessels). 30 Auricular cartilage. 31 Scutiform cartilage. 32–34 Hyoid apparatus (suspending tongue and larynx in floor of throat). 32 Cranial horn of hyoid. 33 Basihyoid bone. 34 Caudal horn of hyoid (thyrohyoid bone). 35 Thyroid cartilage of larynx (forming 'laryngeal prominence' of voice box). 36 Cricoid cartilage of larynx. 37 Cricothyroid membrane. 38 Tracheal cartilages.

Vertebral column, ribs and sternum
39 Wing of atlas (transverse process of cervical vertebra 1). 40 Position of atlantooccipital joint. 41 Spinous process of axis (cervical vertebra 2). 42 Transverse process of axis. 43 Transverse processes of cervical vertebrae. 44 Nuchal ligament. 45 Supraspinous ligament. 46 Spinous process of thoracic vertebra 1. 47 Manubrium of sternum (1st sternebra elongated into base of neck). 48 Xiphoid process of sternum (cartilaginous prolongation of last sternebra into belly wall). 49 Rib 1. 50 Costal arch (fused costal cartilages of ribs 10–12 connected by fibrous tissue with costal cartilage of rib 9). 51 Rib 13 (last or floating rib united with costal arch by fibrous tissue).

Forelimb skeleton
52 Greater tubercle of humerus (point of shoulder). 53 Cranial border of scapula. 54 Cranial angle of scapula. 55 Medial epicondyle of humerus. 56 Medial surface of radial head. 57 Medial collateral elbow ligament. 58 Elbow joint (humeroulnar component). 59 Olecranon process of ulna (point of elbow). 60 Styloid process of radius. 61 Antebrachiocarpal joint (main component of carpal joint).

Hindlimb skeleton
62 Pubic pecten. 63 Cranial pubic ligament. 64 Ischiatic tuberosity (point of buttock). 65 Ischiatic arch. 66 Inguinal ligament. 67 Patella (knee cap). 68 Medial ridge of femoral trochlea. 69 Femoropatellar component of stifle joint. 70 Medial femoral condyle. 71 Medial tibial condyle. 72 Medial collateral ligament of stifle joint. 73 Femorotibial component of stifle joint. 74 Patellar tendon (continuation onto tibia of quadriceps femoris tendon). 75 Fat pad underlying patellar tendon. 76 Tibial tuberosity (insertion of patellar tendon). 77 Cranial border of tibia (tibial crest). 78 Subcutaneous medial surface of tibial shaft. 79 Medial malleolus of tibia. 80 Crurotarsal joint (main component of tarsal joint). 81 Calcaneal tuberosity (point of hock). 82 Metatarsal bone 1 (remains of).

Muscles
83 Temporal muscle. 84 Masseter muscle. 85 Mylohyoid muscle. 86 Sternohyoid muscle. 87 Sternooccipital muscle. 88 Brachiocephalic muscle. 89 Triceps brachii muscle. 90 Flexor muscles of carpus and digits. 91 Supraspinatus muscle. 92 Superficial pectoral muscle. 93 Deep pectoral muscle. 94 Latissimus dorsi muscle. 95 Sartorius muscle. 96 Quadriceps femoris muscle. 97 Pectineus muscle. 98 Gracilis and adductor muscles.

Blood vessels and lymph nodes
99 Angular vein of eye (continuation of facial vein into orbit to connect with ophthalmic venous plexus). 100 Facial vein. 101 External jugular vein (in jugular groove). 102 Femoral artery. 103 Parotid lymph node (small, beneath parotid salivary gland). 104 Mandibular lymph nodes (prominent and normally palpable). 105 Medial retropharyngeal lymph node (large, deeply positioned and up to 5 cm long). 106 Superficial cervical lymph nodes (prominent, palpable if enlarged). 107 Axillary lymph node (related to rib 2, palpable if enlarged). 108 Accessory axillary lymph node (only occasionally present). 109 Superficial inguinal lymph node (often paired, palpable normally). 110 Popliteal lymph node (up to 2 cm long in popliteal fossa and normally palpable).

Nerves
111 Infraorbital nerve (from maxillary component of trigeminal nerve [5] ramifying on side of muzzle).

Glands
112 Parotid salivary gland (diffuse gland moulded around concha of auricular cartilage). 113 Parotid salivary gland duct (crossing surface of masseter muscle). 114 Mandibular salivary gland. 115 Thyroid gland.

Internal viscera
116 Heart. 117 Diaphragm (cranial extent of dome). 118 Kidneys. 119 Urinary bladder. 120 Ovaries. 121 Spleen. 122 Descending duodenum. 123 Caecum. 124 Descending colon.

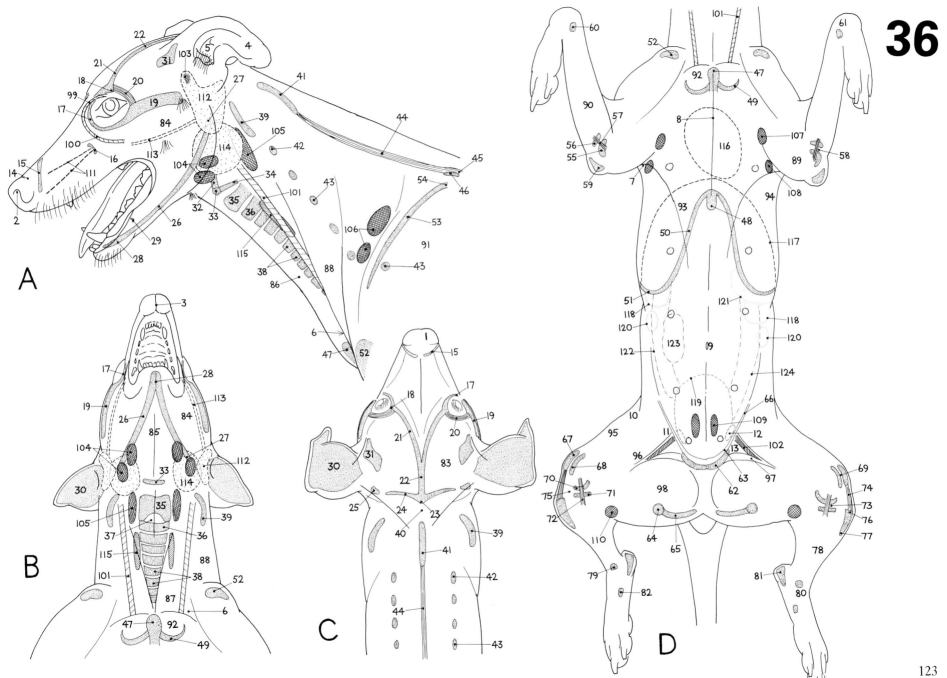

36

123

37

SURFACE ANATOMY OF THE DOG – NECK, THORAX AND FORELIMB

This drawing, and figs 36 & 38, attempt to portray most of the structures that are clearly related to the body surface, either palpable or capable of visualization in surface projection.

Surface features

1 Nasal plane (pigmented hairless skin). **2** External nostril (leading into nasal vestibule surrounded by nasal cartilages). **3** Philtrum. **4** Pinna (visible part of external ear based on auricular cartilage). **5** Jugular fossa (triangular depression at base of neck bordered by sternocephalic, brachiocephalic and superficial pectoral muscles). **6** Axilla ('armpit' between muscles of arm and muscles of chest wall). **7** Median pectoral groove. **8** Tricipital margin of arm (caudal margin of arm based on long head of triceps muscle). **9** Cubital fossa (depression on flexor surface of elbow joint). **10** Ventral boundary of epaxial musculature (thoracic iliocostal muscle attaching onto ribs).

Bones, joints and ligaments

11 Nasal process of incisive bone (bordering bony nasal opening leading into nasal cavity proper). **12** Bony nasal aperture. **13** Infraorbital foramen (passage for infraorbital branches of maxillary artery and nerve). **14** Bony orbital margin. **15** Zygomatic (supraorbital) process of frontal bone. **16** Zygomatic arch (bar of bone connecting face and cranium below eye). **17** Orbital ligament (joining frontal bone and zygomatic arch and completing orbital rim). **18** Temporal line (rostral divergence of external sagittal crest). **19-20** Mandible (lower jaw). **19** Mandibular body. **20** Mandibular symphysis (fibrocartilaginous intermandibular joint). **21** Tracheal cartilages (approximately 35 linked by annular ligaments). **22** Transverse processes of cervical vertebrae 6 and 7. **23** Spinous process of thoracic vertebra 1. **24** Spinous process of thoracic vertebra 13. **25** Nuchal ligament (continuation of supraspinous ligament from summit of first thoracic spinous process to spine of axis vertebra). **26** Supraspinous ligament linking summits of spinous processes of trunk vertebrae. **27–29** Sternum. **27** Manubrium of sternum (1st sternebra elongated into base of neck). **28** Body of sternum (8 sternal segments linked by intersternebral cartilages). **29** Xiphoid process of sternum (cartilaginous prolongation of last sternebra into belly wall). **30** Rib 1. **31** Thoracic inlet (bordered by first thoracic vertebra, sternal manubrium and first pair of ribs). **32** Rib 6. **33** Costal arch (fused costal cartilages of ribs 10–12 connected by fibrous tissue with costal cartilage of rib 9). **34** Rib 13 (last or floating rib united with costal arch by fibrous tissue). **35** Dorsal (vertebral) border of scapula. **36** Cranial angle of scapula. **37** Cranial border of scapula. **38** Caudal angle of scapula. **39** Spine of scapula. **40** Acromion process of scapula. **41** Position of shoulder joint. **42** Greater tuberosity of humerus (point of shoulder). **43** Crest of greater tuberosity of humerus. **44** Deltoid tuberosity of humerus. **45** Lesser tuberosity of humerus. **46** Intertubercular (bicipital) groove of humerus (for passage of biceps tendon surrounded by bicipital bursa and held in place by transverse humeral retinaculum). **47** Medial humeral condyle. **48** Medial humeral epicondyle (flexor epicondyle). **49** Lateral humeral condyle. **50** Lateral humeral epicondyle (extensor epicondyle). **51** Radial head. **52** Lateral radial tuberosity. **53** Medial edge of trochlear notch of ulna. **54** Medial collateral elbow ligament. **55** Lateral collateral elbow ligament. **56** Olecranon process of ulna (point of elbow). **57** Position of elbow joint. **58** Lateral styloid process of ulna. **59** Subcutaneous medial surface of radius. **60** Medial styloid process of radius. **61** Accessory carpal bone. **62** Medial collateral carpal ligament. **63** Lateral collateral carpal ligament. **64** Lateral surface of base of metacarpal bone 5. **65** Position of antebrachiocarpal joint (main component of composite carpal joint).

Muscles

66 Sternooccipital muscle. **67** Brachiocephalic muscle. **68** Supraspinous muscle. **69** Deltoid muscle. **70** Long head of triceps muscle (tricipital margin of arm). **71** Brachial muscle. **72** Biceps tendon (in intertubercular groove). **73** Biceps brachii muscle. **74** Superficial pectoral muscle. **75** Deep pectoral muscle. **76** Radial carpal extensor muscle. **77** Pronator teres muscle. **78** Extensor muscles of carpus and digits. **79** Flexor muscles of carpus and digits. **80** Tendon of ulnar carpal flexor muscle. **81** Tendon of lateral ulnar muscle. **82** Latissimus dorsi muscle. **83** Epaxial muscles (iliocostal muscle forms lateral component of group).

Vessels and lymph nodes

84 Angular vein of eye (continuation of facial vein into orbit where it anastomoses with external ophthalmic plexus). **85** Facial vein. **86** External jugular vein. **87** Omobrachial vein (linking cephalic and external jugular vein across point of shoulder). **88** Cephalic vein. **89** Median cubital vein (entering cubital fossa to join cephalic vein with brachial vein). **90** Accessory cephalic vein. **91** Axillobrachial vein (connection of cephalic vein with brachial/axillary vein in armpit). **92** Brachial artery (pulse may be taken from brachial in cubital fossa). **93** Median artery (enters paw through carpal canal). **94** Mandibular lymph nodes (normally 2 or 3, palpable at angle of jaw). **95** Superficial cervical lymph nodes (normally 2, palpable if enlarged).

Nerves

96 Radial nerve, muscular branches. **97** Radial nerve, superficial branch (palpable on surface of brachial muscle). **98** Cranial superficial antebrachial nerve, medial and lateral branches. **99** Median nerve. **100** Ulnar nerve (palpable crossing medial surface of olecranon of ulna). **101** Infraorbital nerve.

Internal organs

102 Heart (in surface projection extending between ribs 3 and 6, overlapping into intercostal spaces 2 and 6). **103** Base of heart (in transverse plane of rib 3/intercostal space 3). **104** Apex of heart (from which heart beat may be detected at lower end of intercostal space 6). **105** Diaphragm. **106** Basal border of lung. **107** Costodiaphragmatic line of pleural reflection. **108** Parotid salivary gland (diffuse and moulded around concha of auricular cartilage). **109** Parotid salivary gland duct. **110** Mandibular salivary gland (oval, just caudal to angle of jaw).

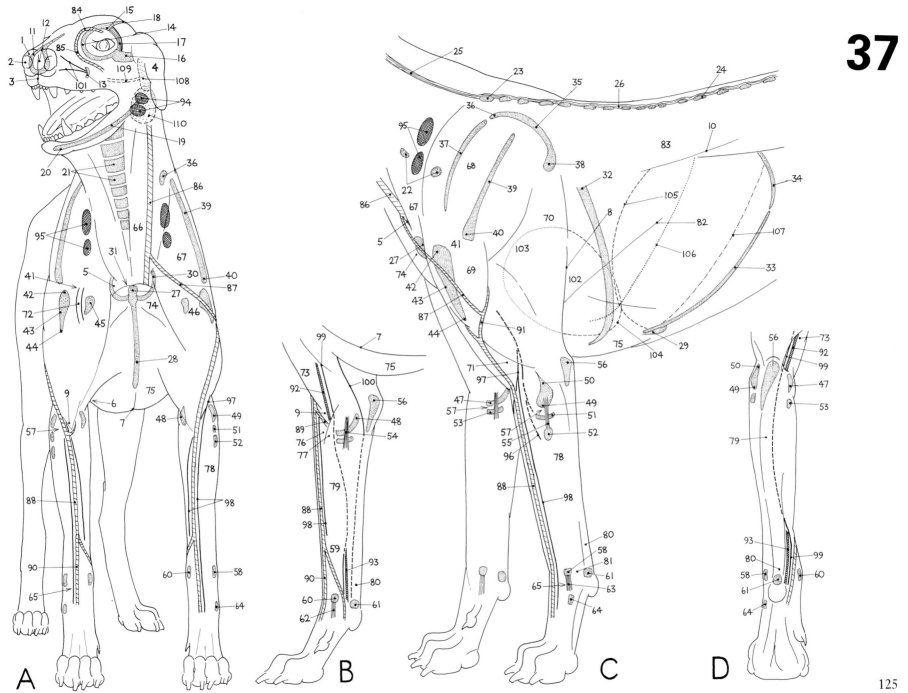

37

125

38

SURFACE ANATOMY OF THE DOG – PELVIS AND HINDLIMB

This drawing, and the previous two (figs 36 & 37), attempt to portray most of the structures that are clearly related to the body surface, either palpable or capable of visualization in surface projection.

Surface features
1 Ischiorectal fossa (lateral to root of tail where bordered by pelvic diaphragm, medial to rump muscles and sacrotuberous ligament). 2 Fold of flank (skin fold joining flank and thigh proximal to stifle joint). 3 Prepuce (sheath or foreskin protecting glans penis and providing a reserve fold of skin for erect penis). 4 Preputial opening (leading into preputial cavity around glans penis). 5 Anus (ducts of paranal sacs open on either side). 6 Circumanal skin (sparsely haired and studded with pores of circumanal glands – modified sebaceous glands producing odoriferous material attractive to other dogs). 7 Penile (urethral) bulb (expansion of spongy erectile tissue below anus and between divergent crura of penile root). 8 Penile crus (fibrous erectile tissue [cavernous body] surrounded by an ischiocavernosus muscle and attached to ischiatic arch – pair of crura attach penis to pelvis and

form basis of penile root). 9 Scrotum (sparsely haired, thin skin). 10 Scrotal raphe (denoting internal subdivision of scrotum into two testicular compartments). 11 Popliteal fossa (depression on flexor surface of stifle joint bounded by diverging hamstring muscles and gastrocnemius muscles).

Bones, joints and ligaments
12 Crest of ilium. 13 Sacral tuberosity (point of croup). 14 Coxal tuberosity (point of haunch). 15 Median sacral crest. 16 Sacrotuberous ligament (extending from ischiatic tuberosity to lateral sacral crest and transverse process of first caudal vertebra). 17 Ischiatic tuberosity (point of buttock). 18 Ischiatic arch (joining ischiatic tuberosities of either side). 19 Greater trochanter of femur (roughly on same level as head of femur and is pointer to position of hip joint). 20 Position of hip joint (joint itself is not palpable). 21 Lateral ridge of femoral trochlea. 22 Medial ridge of femoral trochlea. 23 Patella ('knee cap' – sesamoid bone in tendon of insertion of quadriceps femoris muscle located in femoral trochlea). 24 Position of femoropatellar component of stifle joint. 25 Patellar tendon (tendon of insertion of quadriceps femoris muscle, major stifle extensor muscle). 26 Medial condyle of femur. 27 Medial epicondyle of femur. 28 Lateral condyle of femur. 29 Lateral epicondyle of femur. 30 Lateral fabella (sesamoid bone in tendon of origin of lateral head of gastrocnemius muscle). 31 Medial condyle of tibia. 32 Lateral condyle of tibia. 33 Lateral collateral stifle ligament. 34 Medial collateral

stifle ligament. 35 Position of femorotibial component of stifle joint. 36 Infrapatellar fat pad underlying patellar tendon. 37 Tibial tuberosity. 38 Cranial border of tibia. 39 Subcutaneous medial surface of tibia. 40 Head of fibula. 41 Lateral malleolus of fibula. 42 Medial malleolus of tibia. 43 Calcaneal tuberosity (point of hock). 44 Metatarsal bone 1. 45 Lateral surface of base of metatarsal bone 5. 46 Lateral collateral tarsal ligament. 47 Medial collateral tarsal ligament. 48 Position of crurotarsal component of hock joint (trochlea of talus is palpable). 49 Os penis (penile bone in glans penis formed from ossification in cavernous bodies – maintains glans penis in a permanently 'erect' state. Longitudinal groove on underside of bone houses urethra surrounded by spongy erectile tissue).

Muscles
50 Middle and superficial gluteal muscles (rump muscles). 51 Cranial margin of thigh based on sartorius muscle. 52 Biceps femoris component of hamstring muscles. 53 Semitendinosus component of hamstring muscles. 54 Gracilis muscle. 55 Quadriceps femoris muscle (beneath lateral femoral fascia). 56 Calf muscles (gastrocnemius and superficial digital flexor muscles). 57 Cranial tibial muscle. 58 Tendon of origin of long digital extensor muscle. 59 Tendon of deep digital flexor muscle. 60 Common calcaneal tendon (composite structure formed from Achilles' tendon, superficial digital flexor tendon and tarsal tendons from hamstring and gracilis muscles). 61 Superficial digital flexor tendon.

Vessels and lymph nodes
62 Medial saphenous vein. 63 Lateral saphenous vein. 64 Saphenous artery, caudal branch. 65 Cranial tibial artery (continuation of popliteal artery attaining front of crus by passing through interosseous space between tibia and fibula). 66 Dorsal pedal artery (continuation of cranial tibial artery across flexor surface of tarsus in company with tendon of long digital extensor muscle). 67 Perforating metatarsal artery (continuing dorsal pedal artery between metatarsal bones 2 and 3 into plantar surface of paw). 68 Popliteal lymph node (single prominent node in popliteal fossa, normally palpable).

Nerves
69 Ischiatic nerve (direct continuation of lumbosacral trunk leaving pelvis through lesser ischiatic notch in lateral wall of ischiorectal fossa). 70 Common peroneal nerve (crosses lateral head of gastrocnemius muscle beneath crural fascia). 71 Superficial peroneal nerve. 72 Deep peroneal nerve. 73 Tibial nerve (palpable in crus between common calcaneal tendon and deep digital flexor muscle).

Glands
74 Anal (paranal) sac (paired sacs sandwiched between external and internal anal sphincters – receive and temporarily store secretion from glands in its wall which is added to faecal surface when voiding occurs). 75 Opening of duct of anal sac.

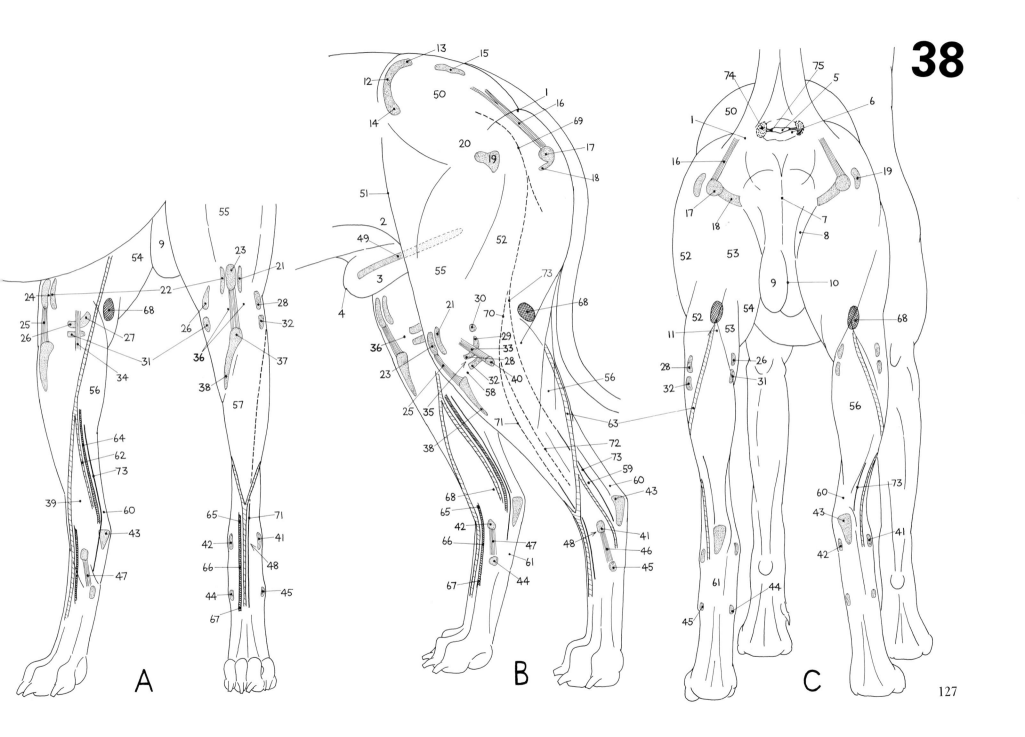

38

127

BIBLIOGRAPHY

Adams D.R. (1986) *Canine Anatomy A systemic Study*. Ames: Iowa State University Press.

Baum H. & Zietzschmann O. (1936) *Handbuch der Anatomie des Hundes*. 2nd edition. Berlin: Paul Parey.

Bourdelle E. & Bressou C. (1953) *Anatomie Régionale des Animaux Domestiques*. Vol IV – Carnivores: *Chien et Chat. Paris: J-B Baillière.*

Boyd J.S. & Patterson C. (1991) *A Colour Atlas of Clinical Anatomy of the Dog and Cat*. London: Wolfe.

Bradley O.C. (1959) *Topographical Anatomy of the Dog*. 6th edition revised by T.Grahame. Edinburgh: Oliver & Boyd.

Budras K-D. & Fricke W. (1991) *Atlas der Anatomie des Hundes. Lehrbuch fur Tierärtze und Studierende*. 3rd edition. Hannover: *Schlütersche.*

Done S.H. Goody P.C. Evans S.A. & Stickland N.C. (1996) *Colour Atlas of Veterinary Anatomy. Volume 3 – The Dog and Cat*. London: Mosby-Wolfe.

Dyce K.M. Sack W.O. & Wensing C.J.G. (1987) *Textbook of Veterinary Anatomy*. Philadelphia: W B Saunders.

Ellenberger W. Dittrich H. & Baum H. (1956) *An Atlas of Animal Anatomy for Artists*. 2nd edition revised by L.S.Brown. New York: Dover.

Evans H.E. & Christensen G.C. (1979) *Miller's Anatomy of the Dog*. 2nd edition. Philadelphia: W B Saunders.

Evans H.E. & de Lahunta A. (1988) *Miller's Guide to the Dissection of the Dog*. 3rd edition. Philadelphia: W B Saunders.

Sisson S. & Grossman J.D. (1975) *The Anatomy of the Domestic Animals*. Vol 2. 5th edition edited by R.Getty. Philadelphia: W B Saunders.

Taylor J.A. (1955–1970) *Regional and Applied Anatomy of the Domestic Animals*. Parts I – III. Edinburgh: Oliver & Boyd.